PRERATIONAL INTELLIGENCE:
INTERDISCIPLINARY PERSPECTIVES ON THE
BEHAVIOR OF NATURAL AND
ARTIFICIAL SYSTEMS

STUDIES IN COGNITIVE SYSTEMS

VOLUME 26/3

PRERATIONAL INTELLIGENCE:
INTERDISCIPLINARY
PERSPECTIVES ON THE
BEHAVIOR OF NATURAL AND
ARTIFICIAL SYSTEMS

Edited by

Jeffrey Dean

Cleveland State University, Cleveland, Ohio, U.S.A.

Helge Ritter

University of Bielefeld, Bielefeld, Germany

and

Holk Cruse

University of Bielefeld, Bielefeld, Germany

Springer-Science+Business Media, B.V.

A C.I.P. Catalogue record for this book is available from the Library of Congress.

ISBN 978-94-010-3792-1   ISBN 978-94-010-0870-9 (eBook)
DOI 10.1007/978-94-010-0870-9

Printed with the support of the University of Bielefeld, Germany
Gedruckt mit Unterstützung der Universität Bielefeld

TABLE OF CONTENTS

PREFACE

"Intelligence" and "intelligent" are an essential part of being human. As such they are a subject of interest for many different academic disciplines. Although the concept is used continuously in everyday life, when intelligence is examined more closely, it turns out to be multi-facetted and extremely resistant to precise definition. In its colloquial sense, intelligence is equated with acquiring knowledge and using it in ways that further the purposes of the individual. Novelty is a primary attribute: intelligence is evident in acquiring "new" knowledge, making "new" associations of previously acquired knowledge, and behaving adequately in "new" situations (cf. Webster's Ninth New Collegiate Dictionary, *1: the ability to learn or understand or to deal with new or trying situations*).

Intelligence, like many specifically human attributes, appears to be closely tied to language and the social interactions it allows, to logical thought and to reason (cf. *1: continued: the skilled use of reason, and 2: the ability to apply knowledge to manipulate or interpret one's environment or to think abstractly*). Both explicitly and implicitly in the references to understanding, knowledge and abstract thinking, this definition identifies intelligence with rational thought. This definition fits with our introspective notion of how humans function as intelligent beings. Most of the time most of us feel ourselves to be a unified personality acting with one voice, observing ourselves and the world around us, storing information about what we observe, weighing alternative plans and then selecting one or more actions for performance. Of course, we know intellectually that activity in many parts of the brain eludes our introspection. These include sensory stations on the way from receptor activation to our conscious experience, motor stations converting our will to act into patterns of muscle activation, and many centers affecting our emotions. Our introspection also does not have access to the processes themselves. What happens when we view a Necker cube, learn a new fact or decide to move? What goes on in various limbic centers when our mood changes. Neuroscience tells us that neural processing is often distributed among separate anatomical centers, each apparently specialized for different subtasks, but we feel that the outcome is unified, giving our personality a single point of view and a single actor. Furthermore, we feel that our intelligence reflects the rational processes of this actor, processes that can be

J. Dean et al. (eds.), Prerational Intelligence: Interdisciplinary Perspectives on the Behavior of Natural and Artificial Systems, vii-xiii.
© 2000 *Kluwer Academic Publishers.*

expressed in symbols and rules of behavior, like those we communicate to our children.

With the development of computers – machines that easily simulate many low-level abilities of human thinking, it was natural for engineers and scientists to adopt as a goal the implementation of this notion of rational intelligence in machines. Human beings remained the measure of intelligence. The test of an intelligent machine was whether a human could distinguish it from another human by interacting with it. Because formal, symbolic language mediates so much of human interactions, an intelligent machine had to be able to interact through language. The way to make one was to provide the machine with information in the form of symbols and to implement a way for the machine to systematically add information in order to build up a large knowledge base culminating in a so-called "world model" of itself, its surroundings and the relationships between them. Finally, the machine needed to be able to operate on this knowledge, whereby operation on knowledge was oriented on logic in the belief that human thought in its true form is rational and can be captured in specific, formal rules. When the machine was probed with a question, it could then operate on its knowledge to interpret the question, resolving any ambiguities by taking the context into account or by asking an appropriate question for clarification, and then respond appropriately, like a human. The underlying conception was that of an intelligent being as a processor of information.

Reflecting this anthropocentric bias, initial attempts focussed on intelligent systems operating wholly in the realm of symbols: questions, responses and stored knowledge were all formulated as streams of symbols. This bias was also true of much thinking in the field of artificial intelligence, as was apparent in the emphasis on knowledge and cognition, on symbolic representation and on logical manipulation of these symbols in formally described, closed systems. The related approach in perception, particularly in the field of machine vision or language interpretation, was to search for basic features and relationships among them that uniquely identify objects or situations in the world. The related approach in motor control was to emphasize precise analytical solutions, usually computed by an omniscient, central controller using exact sensor information to calculate reference signals for appropriate servocontrollers. Symbolic interaction is, of course, a major part of our behavior as human beings, but focussing on it alone ignores the context in which it occurs, the origin of the symbols and the need to act in complex, natural environments at biologically appropriate speeds.

When this traditional approach was applied to systems expected to behave autonomously – without human supervision and outside protective laboratory environments, weaknesses became apparent. Despite some success in specific areas, particularly in military and related applications where financial constraints were weak, attempts to apply the approach to everyday tasks ran into problems both with the implementation of the concept and with the concept itself. Besides problems of complexity and inadequacies in the world models and operations on those models, unanticipated problems arose in areas assumed to be trivial – namely preparing inputs to the model and carrying out the selected actions. Early simulations had relied on humans to perform the necessary preprocessing to turn raw sensor data into symbols. When machines were asked to do this work for themselves and confronted with noisy sensor signals and incomplete or ambiguous data, they failed. Interpreting a visual scene from the activity induced in arrays of photosensors or deciphering a sentence from the signals of a microphone proved to be immense computational problems. Simply making faster computers did not make all the problems disappear or generally lead to more satisfactory results. Similarly, attempts to control real robots with multiple actuators and sensors, which may be noisy, variable and unreliable, ran into similar computational limits and produced results far from the graceful, apparently effortless movements of animals.

These failures indicate that an intelligence of process, if not of content, is necessary in the stages before and after any rational processor, namely in the stages that extract information about the world and transform logical decisions into actions. For an agent in the real world, intelligence must be defined in a broader sense, without reference to a specific mechanism, as the ability of a system to function in new situations in ways that are useful to the agent or to the larger entity of which it is a part. In many situations this ability entails capacities for generalization, flexibility, plasticity and learning, and robustness – attributes that will be considered in more detail in the course of the book. It is this intelligence of process in the absence of symbols and logical reasoning that we call *Prerational Intelligence*.

Animals, which by nature are autonomous, provide models for intelligent behavior and, in the last analysis, all that we identify with human intelligence must be implemented using the physiological and anatomical mechanisms we share with other animals. Increasing knowledge about mechanisms controlling animal behavior and newer results with artificial systems point to an alternative approach to the creation of systems that are capable of intelligent behavior in the sense of the last paragraph. The premise of this approach, which will be explored in the following chapters, is that traditional notions of intelligent behavior, which formed the basis for much work in artificial intelligence and cognitive science, presuppose many basic capabilities that are not trivial and that these capabilities may be best understood as emerging from interaction and cooperation in systems of simple agents, where agents can be neurons, networks, individuals or any identifiable unit capable of acting on and being acted upon by its environment. These mechanisms, we presume, are the basic mechanisms of prerational intelligence.

In such systems, knowledge may not be explicitly represented. Storage and processing of experience may be distributed over many elements, so it is robust against loss of a few elements and degradation of performance is graceful. Processing can occur in a parallel manner, without a single master processor. Because processing is distributed, it may not lead to a single answer but instead be inexact or fuzzy. This fuzziness in turn can lead to adaptive generalization and provide the opportunity for self-improvement through comparison of actual performance with some evaluation thereof. Most important, intelligent behavior is seen to emerge from the dynamic properties of the whole system and these systems themselves, whether groups of individuals, the individuals themselves or modules within an individual, are part of a dynamic interaction with their surroundings.

Because these properties are evident in animals, working examples of intelligent autonomous systems, where they apparently perform the work of processing sensory inputs and converting decisions into actions, we propose to refer to this intelligence of process as prerational intelligence. Evolution has provided animals with the appropriate mechanisms. Recent work in various simulations and artificial life indicates that analogous mechanisms can be successfully adapted to machines and computers, where the intelligence of process arises by design. Thus, these intelligent systems are prerational in that the mechanisms arose in the course of evolution in animals where one would be reluctant to speak of rational thought and in the sense that they handle the inputs and outputs to what appear to us to be reason and intelligence in humans. However, the example of human intelligence, which of course must also rest on biological mechanisms common to other animals, indicates that rational intelligence in its highest sense must also reflect the workings of a dynamic system, one realized in a brain with an enormous number of elements and interconnections.

The question to be addressed in the book then is how this paradigm of prerational intelligence arising in systems of interacting agents can be reconciled with experimental findings and theoretical approaches in the different academic fields that consider intelligence and intelligent behavior.

The present book is the final product of a year-long Research Group entitled "Prerational Intelligence" that was held in Bielefeld at the Center for interdisciplinary Studies. The organizers created the name and chose the theme for the reasons indicated above. To repeat, the premise explored by the group is that many basic capabilities necessary for autonomous intelligent behavior, and maybe even rational intelligence itself, are best understood as phenomena emerging in systems of simple agents. The term "prerational intelligence" was chosen to designate this intelligence of process in the absence of symbols and logical reasoning; it is the subject of the book.

The main impetus for choosing this theme came from the organizers' research, which focuses on the way animals and artificial systems utilize information about their surroundings in order to move and act adaptively. The analysis of the collective properties of systems of interacting agents, however, is a problem that occurs repeatedly in many disciplines. For this reason, the ZiF Research Group included representatives from many different fields concerned with the question of intelligent behavior. Biology, psychology, philosophy, mathematics, computer science, robotics, engineering, and artificial intelligence were all represented in the core group in residence at ZiF.

In the course of a year, the group considered the many facets of intelligent behavior and discussed approaches to understanding biological intelligence and duplicating it in artificial systems. Presentations of the many speakers invited to the conferences hosted by the Research Group, most of which are contained in volumes 1 and 2, contributed significantly to the work of the group. As the year drew to a close, the participants felt the need to summarize the results of their deliberations.

All participants were confident that promising outlines for an account of natural intelligence are taking shape. Major holes do remain to be filled and it clearly is too early for all disciplines to agree on a single view of intelligence. Therefore, the Research Group decided that until a unified theory of intelligent behavior crystalizes, one that can be translated into every discipline, the most appropriate format for a report is to let the representatives of each discipline present their own review of the concept of prerational intelligence and its relevance to questions in their respective disciplines. Hence, the task for the contributors was to review the state of progress, to focus on questions and findings in their fields they deem important and, in particular, to consider interdisciplinary issues and implications of research in other areas.

The book begins with chapters from biological and psychological perspectives, because these two fields consider working examples of intelligent systems. *Biological Perspectives on Prerational Intelligence* identifies features of intelligence common to humans and animals and then examines the features of nervous systems and control strategies that appear relevant. It concludes with a discussion of the relationship between network architectures and intelligent behavior and a possible extension to rational processes. *Prerational Intelligence from the Perspective of Psychology* reviews the concept of intelligence in psychological research and theory and then considers the implications of systems exhibiting prerational intelligence for understanding human behavior. In particular, it compares and contrasts this approach with that of behaviorism and with the developmental theories of Piaget. The next chapter, *Prerational Intelligence from the Perspectives of Robotics and Engineering,* turns to artificial systems, and asks the question to what extent the paradigm of prerational intelligence can substitute for or improve on classical algorithmic approaches to robot design and control, given the potential economic and application constraints. From the perspective of engineering, the chapter examines prerational intelligence in the context of control techniques; it considers the possible utility of prerational approaches to synthesizing complex intelligent behavior given the different constraints of natural and artificial systems. Specific topics include models for control architectures and modular approaches to synthesizing complex intelligent behavior from elementary behaviors. The next chapter, *Computer Science Perspectives on Prerational Intelligence,* addresses similar issues in the realm of computer science, focussing on appropriate ways to code, store and retrieve data and possible uses of biologically inspired algorithms to evolve and optimize intelligent systems. It also treats formal issues of complexity as related to intelligence. The *Mathematical Perspectives on Prerational Intelligence*, rather than examining issues raised by specific applications, consider more general questions related to prerational intelligence, including measures of complexity and algorithms for searches, for pattern formation and recognition and for generation of structures. A particular focus is the theory of dynamic systems as applied to animals and artificial systems. The final chapter, *Philosophical Perspectives on Representation, Goals, and Cognition,* examines from a philosophical standpoint several key concepts.

For this book, each contributor was asked to reflect from the point of view of a given discipline on the nature of intelligence and intelligent behavior and on the general principles of process and organization that might lead to their appearance. Several key questions are the following. What is meant by intelligence? What actions by an animal or a machine would we consider to show unequivocal intelligence? If, as we will argue, intelligence can be seen as a continuum in a multidimensional space, what features are useful for ordering degrees or kinds of intelligence? What mechanisms can lead to intelligent behavior? Are they necessary or merely sufficient? Finally, and perhaps most important, what can different disciplines learn from each other and are there general principles that are valid across disciplines?

ACKNOWLEDGMENTS

In closing, the editors express their deepest appreciation to the directors of the ZiF for their support of the Prerational Intelligence Research Group, to the entire staff of the ZiF for all their technical support facilitating the work of the group and the conferences and to the Universität of Bielefeld and the Land Nordrhein-Westfalen for making the ZiF the unique working environment that it is. We thank Patrick Ziemeck for his excellent technical support, and, above all, we express our gratitude to Lilo Jegerlehner for her competence and tireless effort in preparing the manuscripts for publication. With sadness the editors also wish to note the tragic early death of Peter Lanz, a member of the Research Group and major contributor to its work.

Universität Bielefeld   *Jeffrey Dean, Holk Cruse, and Helge Ritter*

JEFFREY DEAN*, HOLK CRUSE**, and HELGE RITTER**

INTRODUCTION TO VOLUME 3

1. INTELLIGENCE, INTELLIGENT BEHAVIOR AND RELATED STRUCTURAL FEATURES

Early on the organizers and the Research Group decided not to attempt a rigid definition of intelligence, but rather to focus on features commonly associated with the concept of intelligence or intelligent behavior. What do we mean by intelligence? What actions by an animal or a machine would we consider to show true intelligence? If, as we will argue, intelligence can be seen as a continuum, what features are useful for ordering degrees or kinds of intelligence?

If we can agree on a working definition of intelligence, then it is a simple matter to define prerational intelligence as intelligence that does not rely on rational thought, symbolic representations and logical reasoning. The ultimate goal is to consider whether there are common organizational principles in systems showing such behavior.

Standard dictionary definitions of intelligence highlight the ability to deal with new or trying situations but, at the same time, they link intelligence strongly to rational thought. The viewpoint adopted here is that this linkage represents an anthropocentric bias emphasizing highly advanced human capabilities. This anthropocentric bias influenced much early thinking about artificial intelligence, leading to a focus on knowledge and cognition, on symbolic representation and on logical manipulation of these symbols in formally described, closed systems. Similarly, control of movement relied on exact analytical solutions to provide reference signals for servocontrollers, while perceptual processing attempted to interpret sensor data by applying stored knowledge. However, a truly complete explanation and simulation of these rational abilities must extend to the axioms, such as the origin of the symbols and the logical rules. Consideration of the foundations for these higher functions from evolutionary and neurophysiological perspectives requires a different approach, one suggesting that many behaviors that humans naturally describe as logical processes in fact are realized in biological systems using quite different mechanisms.

1

J. Dean et al. (eds.), Prerational Intelligence: Interdisciplinary Perspectives on the Behavior of Natural and Artificial Systems, 1–10.
© 2000 *Kluwer Academic Publishers*.

Thus, the Research Group places less emphasis on the use of reason and symbols, and more on the function. As a starting point, it characterizes intelligence as "the ability to deal with new situations". This is the nominal use of intelligence; it clearly is an attribute of a behaving agent and most clearly relates to action selection. In addition, there are the adjectival and adverbial uses of intelligent: usually with an implicit value judgement, about the characteristics and mode of operation of a particular behavior, action, design or process. The Research Group feels that the latter use can profitably be extended to modules or subprocesses below the level of an autonomous agent, such as, for example, perceptual processes that identify features in the environment or motor processes organizing movements.

To begin and provide a common reference for the chapters that follow, we will briefly consider what features might characterize intelligence. These points are revisited in the *Biological Perspectives* and individually in other chapters.

First of all, it is important to emphasize that intelligence is an attribute of an agent acting in a context. Our judgements of intelligence are affected by aspects of both the agent and the context. Intelligence resides in the relation between the set of behaviors or outputs that an agent could produce in a given context and the set it actually produces. As an attempt to provide axes for the space of intelligence, we suggest the following six attributes characterizing the relationship of agent and context.

Autonomy: An essential, if trivial, part of the definition is that the agent perform on its own, utilizing external inputs as needed but without external guidance. More generally, one might insist that intelligent processes be of utility in maintaining the autonomy of the agent, or when agents in a collective are under consideration, that of the larger unit (e.g., organism, society) of which they are a part.

Complexity in the task, expressed either in the number of relevant factors (e.g., goals, variables and value criteria) or the non-linearity of the relationship between inputs and adaptive outputs: Complexity in the task need not correlate with intelligence in the agent: hard-wired solutions are usually not seen as intelligent. However, the adverbial use of intelligence, performing intelligently, can be extended to include hard-wired solutions produced through evolution. Similarly, complexity in the agent's procedure is usually not regarded as intelligent if equivalent, simpler procedures could be used.

Generalization is necessary to respond adequately to novel situations: implementing a continuous transformation from a continuous input to a continuous output would represent very simple generalization (low intelligence). Conditional generalization, restricting generalization to appropriate input domains, would represent a higher level of intelligence that would be equivalent to category building. Generalizing into qualitatively new domains, if accompanied by appropriate controls and evaluations, would represent a still higher level of intelligence and contribute to generating new behaviors and making the system open. Insight is a high form of the latter.

Flexibility in the agent, both with respect to varying inputs and to constant inputs: Here again, as in the case of generalization, implementing a continuous transformation from a continuous input to a continuous output would represent simple flexibility (low intelligence). Discontinuous transformations, characterized by qualitatively different strategies or output patterns, would represent greater flexibility. At the process level, these discontinuous changes may require discontinuities in underlying control systems, that is, reorganization or recruitment of new elements. For example, flexibility in response to a constant input is needed to provide alternative behaviors for an adequate response to initial failure. Stochastic variability would be a low-level flexibility, which still could be evaluated for its efficiency. Learning and memory open a path to greater flexibility. Persistence in an ineffective behavior indicates lack of intelligence.

Adaptive plasticity and learning by the agent to adapt to changing conditions based on past experience: Parameter adjustment in a predefined space, as in self-calibration of sensors or effectors, would be a low-level competence. Formation of new categories for perception, behavior or motor skills would be higher level. Reflexivity, permitting self-generated categories themselves to be objects for building new categories, would be a still higher form. A prerequisite for learning is that the system have some failure tolerance. Memory, in some form, can provide an improved ability to avoid persistence in ineffective behaviors.

Intentionality, by which is meant directed, adaptive selectivity in the agent's interactions with the environment and in its procedures: Two examples would be, first, identification of and selective attention to behaviorally relevant inputs and, second, selectivity in the choice of behavior, that is, in the choice of movement or more generally of perception-action patterns, based on internal states and external inputs. Brute force approaches based on exhaustive, undirected search of the potential action space, as opposed to selective search of the space of likely solutions, would correspond to low intelligence. Active perception and exploration represent higher levels of intentionality.

Several derivative attributes become important in judging the intelligence of a behavior, particularly in competitive situations or in practical implementations. These include: 1) utility, as a measure of the relationship between costs and benefits, possibly defined with respect to implicit or explicit goals or purposes, 2) efficiency in performing a behavior or achieving goals, including parameters like speed, 3) robustness, in the sense of insensitivity to noisy inputs and outputs, 4) open-endedness in the domain of action, as a result of self-reflective or recurrent processes, as opposed to an ability limited to a predefined, circumscribed set of contexts, and 5) anticipation, the extent to which the agent can rely on a prediction of the outcome of an action rather than feedback signals related to consequences.

Based on a superficial inspection of these attributes, several structural features can be identified as possible prerequisites for a system to produce intelligent behavior. Complexity and discontinuous control spaces may necessitate modularity in the control structure, both horizontally in partitioning tasks at similar levels and vertically in providing a hierarchy of control. Flexibility and robustness may favor the incorporation of redundancy. To avoid confinement to a predefined context, systems may need to learn and evolve. Anticipation and intentionality may require internal models, that is, representations of the state of the agent and its environment that can be manipulated in the absence of overt behavior.

These hypotheses can be gleaned from considering what we know about the structure and function of nervous systems in animals. At the same time, mere correlation between performance levels observed in animals and the principles we discern in their control structures on the basis of our still very incomplete information does not prove a necessary relationship between the two. Therefore, an important current line of investigation is to test how much intelligent behavior can be produced by systems designed without some or all of these complexities, such as systems lacking explicit goals, representations, and models. Many examples are discussed in the subsequent chapters.

2. COMPARISONS OF ANIMALS AND ARTIFICIAL SYSTEMS

Besides the focus on individual features of intelligent behavior, a second important aspect of the Research Group was the consideration of animals and artificial systems in parallel. The motivation for this decision is the fact that animals apparently have solved many of the sensory, motor and integrative problems that bedevil technical approaches. At a functional level, many of the same tasks must be performed. Many features of intelligent behavior apply to both animals and technological systems. Thus, much may be learned from a better understanding of biological systems for the purpose of comparing and contrasting them with technical systems.

In considering a possible transfer of biological principles to technical systems, it is important to be aware of important differences between animals and artificial systems. At the same time, these differences provide the opportunity to distinguish between what is necessary for achieving a particular level of intelligent behavior and what is merely sufficient.

To conclude this introduction, we would like to point out some of these differences. Perhaps the most important is that technical systems are not autonomous in the sense that animals must be. Beyond this basic difference, several issues are still very much undecided. The starting point for the comparison is the observation that technological systems often appear to have significant advantages in their component sensors, actuators and electronic control systems. Nevertheless, biological systems are quite superior in a variety of functions. Therefore, the general question is whether this superiority is achieved in spite of the inferiority of biological components, which are limited by material constraints and by the process of evolution, or precisely because of the biological materials and the features that appear to be inferior. An answer to this question would indicate what might be gained by emulating biological systems. Several specific features can be considered.

First, biological effectors, the muscles, appear to be unnecessarily complex. They have complicated variable, non-linear combinations of spring-like characteristics and the force they can provide differs according to the speed of contraction. Such characteristics can be duplicated by adding appropriate controllers to motors, but the question is whether anything is gained by the additional computations required. Some evidence suggests that these properties may simplify the overall control process. Clearly the muscle properties are used in some systems to produce movement in the absence of explicit excitatory motor commands, as in the swing movement of some bipeds or the contraction of the extended leech in crawling.

Second, biological sensors are equally complex. They typically have complex frequency responses, which means that they signal both the current value of a parameter and something about its rate of change. Predictive control is well known in technical systems, but is there anything to be learned from the way biological systems do it or is this just one feature that contributes to making a specific, evolved system – an animal – well adapted to a particular niche. In other words, are sensory systems limited by biological mechanisms to perform the way they do? Have they evolved in this direction in order to compensate for characteristics of the other elements, for example, the slow response times of muscles and finite conduction and switching times of neurons? Or is there a general advantage to be gained by emulating these features?

Third, in biological systems the control functions must be realized in the dynamics of neural processes. The multitude of ion channels, governed by electric potentials or chemical transmitters and biochemical mechanisms and modulated with both short and long time constants, make individual neurons exquisitely plastic computing units, despite their disadvantage in speed with respect to electronic circuits. Excitatory connections between neurons can be used to create positive feedback to stabilize a given pattern of activity or to organize groups of neurons into cooperating units. Excitatory connections also can be used to establish sequences of activation and transitions between states. Inhibitory interactions have the opposite effect, preventing simultaneous activity, but in complex circuits they can also be combined with temporal plasticity, that is, fatigue or adaptation or rebound from inhibition, to generate patterns of activity.

The complexity of neurons and their interactions together with the relatively slow processing speed means that initiating, specifying and terminating movements require a measurable time. Delays, which are evident in reaction time experiments, may reflect conditional, serial processes, such as when a particular movement or computation must finish before another can begin. However, they may also represent the process of generating a specification within a network that may include parallel as well as serial channels. This process is not yet understood, but it is clear that the nervous system does its computations in some unintuitive ways. Natural selection often places a premium on reaction speed, so it may be important to be able to produce some action even if it is not perfectly specified, rather than waiting until all processes are complete and checked for consistency, as is usually the case in robotics.

Working with neurons also affects the ways features of the environment are encoded and analyzed, or, in a more neutral formulation, the way patterns of sensory activity are transformed to affect behavior and how both input and output patterns may be stored and used to affect subsequent performance or selection of behavior. The question is often formulated as one of how information is represented in the nervous system. In artificial systems, symbolic representation provides a very concise representation of features, parameters, objects or concepts. In animals, these functions must be done with neurons. In higher animals, regular topographic mappings of simple sensory features in arrays of neurons with sensitivities that vary in systematic ways lead to the notion of a spatial representation of stimulus characteristics within the brain. Identification of neurons responding more or less selectively to more complex stimuli, analysable as combinations of simple features, suggests that such neurons or groups of neurons detect that particular complex stimulus. However, how such regular mappings come about and, more important, how the activity is used are still unclear.

Networks of neurons may also restrict the types of computations that can be performed. Although it is possible to show that networks of artificial neurons can approximate arbitrary functions of very general classes, this need not mean that they can be realized with real neurons given constraints on numbers or processing accuracy.

Fourth, animals rarely respond to stimuli in an entirely predictable way.
This means that behavior is not simply triggered by sensory stimuli but it
arises from the interaction between the environment and internal factors that
define the state of the animal. Plasticity in high-level sensorimotor control
may be required by variability in the effectors and sensors as well as for the
purpose of acquiring new movement patterns. Plasticity, under the control
of the high-level mechanisms, in the use of low-level control mechanisms
is evident in reflex modulation, which has been demonstrated for various
rhythmic and non-rhythmic behaviors.

*Fifth, high-level sensorimotor control in biology involves adaptation to
a variety of predictable and unpredictable situations.* Animals meet these
demands with a combination of feed-back and feed-forward mechanisms.
Uncertainty in the environment or in the animal requires that a particular
state be specified to which feedback signals in the form of sensory inputs
or recurrent signals can be compared. Taking the vestibulo-ocular reflex as
an example, the output characteristics of the eye muscles are not so constant that once the reflex gain is calibrated, the response – compensatory
eye movement to maintain eye fixation during head rotation – will ensure
adequate compensation over long times. Therefore, poor calibration is detected based on sensory feedback signals reflecting a disparity in fixation
before and after head rotation as well as the reflex action. Over the short
term, however, the physical relationships among stimulus, rotation of the
head, and reflex response do not change from one time to another, assuming
the torques provided by the eye muscles are known. Predictability of this
kind means that response speed can be increased by omitting the feedback
cycle, with its delays in waiting for sensory feedback, comparing actual and
desired states and generating new motor outputs. Motor activity based either on intrinsic activity or on parameters specified by sensory inputs can
be sufficient. As this example already suggests, animals vary their reliance
on one or the other and the relative importance of the two varies with task
conditions. Control systems for swimming and flying often incorporate a
well-developed central pattern generator, which is consistent with actions
in a uniform, relatively predictable and forgiving medium. Control systems
for walking, on the other hand, apparently place more reliance on peripheral
control mechanisms, because substrates are typically inelastic and unpredictable. In robotics, control strategies have not yet reached this state of
adaptivity.

An important distinction in the field of learning is between procedural
and declarative knowledge. This distinction, too, will be elaborated in later
sections. Here, we will simply note that in focussing on how animals and
machines process inputs in order to behave and act in the real world, we em-

phasize skills that are procedural rather than declarative or cognitive. Our premise is that an important reason for the difficulties in the traditional approach to artificial intelligence is the lack of attention given to just such procedural skills. On the one hand, such skills are inherently necessary steps in going from raw sensory data to the abstract and symbolic representations that are the focus of much work in cognitive science. On the other hand, the cognitive skills themselves, at least in humans and animals, must be implemented in neural circuitry and processes that themselves can be analysed as procedures performed by distributed networks of simple processing units. Thus, we believe that a better understanding of how complex procedures can be performed by such networks will eventually shed light on how higher rational abilities are actually realized.

Sixth, the relationship of robustness and complexity, both of which are characteristic of many biological systems, is still very unclear. The problem can be illustrated using an analogous problem in ecology: are complex ecosystems inherently more stable with respect to environmental change than those with few species? Intuitively, the answer would seem to be yes and yet some experiments suggest that adding species can make simple ecosystems less stable. (Current evidence suggests the complexity up to a point usually improves stability but this point is not nearly as complex as most natural ecosystems.) In high-level motor control, the analogy would be that adding complexity by including additional control mechanisms and features can lead to emergent phenomena, that is, behavior of the whole system that cannot simply be predicted on the basis of the behavior of the components. These emergent phenomena may be either desirable or undesirable with respect to the optimal performance of the whole system; in any event, they make a technical system less transparent for a designer. A critical requirement for many technical systems is a safety evaluation, which is difficult if the behavior cannot be predicted under all conditions.

The general problem is that adding more mechanisms requires a suitable way to combine them. Linear summation, if the influences can be expressed in a common currency, provides one possibility, but then there is the problem of finding proper weightings. Logical selection criteria or sub-symbolic selection procedures are other possibilities. Examples of logical criteria are the if-then constructions in classical expert systems or the subsumption architecture of Brooks, in which one unit interrupts the output of another if certain conditions hold. Linear or non-linear combinations of subnet outputs in artificial neural networks with multiple "experts" are an example of the latter. Perhaps one secret to the success of biological sensorimotor control, in addition to the plasticity of the individual elements, lies in their very distributed and parliamentary-like organization. As in a democratic parlia-

ment, there usually are many different opinions represented besides that of the majority. Similarly, in several neural networks there are elements that apparently are at odds with the overt function of the entire system. Given time or changing circumstances these minority opinions may become the majority, changing the overt action. This complexity also is important for determining the stability of the control in the sense of responding to altered circumstances or intentionally changing from one behavior to another, two aspects at the center of the dynamic systems approach.

Seventh, biological systems arise through evolution whereas artificial systems arise by design. For biological systems, this means that reproductive success, the animal's inclusive fitness, is the overriding goal for each individual. Whether this ultimate factor is explicitly represented in order to directly guide behavior or whether it acts only indirectly through a variety of proximate factors is a matter for experiment. Thus, in using the terminology of goals, one must make the viewpoint clear; some goals may be only in the mind of an observer, not in the control system of the animal (see, in particular, *Philosophical Perspectives*). For designed systems, the same distinction can be made. Moreover, the designer provides any goals, whether they are implicit or explicit. Therefore, they can be quite arbitrary, especially because most artificial systems are not subject to the constraints implicit in full autonomy.

Biological evolution and autonomy require animals to be self-organizing. For simple systems with few elements, the information capacity of the genome suffices to completely specify development, but with increasing complexity the strategy must change to one of providing general rules for development. By using patterns of regularity in the environment, the latter strategy can produce more precise structures than are possible by genetic specification alone, especially when the specific environment that the individual will experience is unpredictable. However, evolution also constrains development to be gradual and continuous: animals expressing a new feature must themselves be fit. New features must build upon and be compatible with existing elements. Radical, global changes are apt to be so disruptive that fitness is low and they fail to survive and reproduce. Both developmental processes over the life of an individual and evolutionary changes in populations over generations can be analyzed using the language of dynamic systems.

3. CONCLUSION

These then are some of the issues addressed by the Research Group in the following chapters. Themes that the Research Group believes contribute significantly to (pre)rational intelligence include the following: 1) that intelligent behavior in biological systems often is generated collectively in systems of simple agents interacting using simple rules, 2) that this organizing principle leads to robustness, 3) that the interaction with the environment, in particular the use of affordances, is an important aspect of intelligent action, 4) that the dynamics of these interactions is an important subject for study, and 5) that biological implementations often involve approximations and simplifications rather than exact logical or analytical solutions.

In summary, as the individual chapters explain, the Research Group as a whole believes that it is sensible to consider intelligence or intelligent behavior as forming a continuum in a multi-dimensional space, part of which is occupied by rational intelligence. Because this rational intelligence has evolved in systems of simple units, neurons, that interact in ways below the level of symbols or logic, it seems reasonable not to expect a sharp discontinuity between rational intelligence and non-rational or prerational intelligence. By studying the latter, we hope to gain insight into mechanisms that may support the former as well as to improve our understanding of abilities that are not so amenable to logical analysis.

*Cleveland State University, Cleveland, Ohio, USA
**University of Bielefeld, Germany

JEFFREY DEAN*, RALF BECKERS**, and HOLK CRUSE***

BIOLOGICAL PERSPECTIVES ON PRERATIONAL INTELLIGENCE

1. INTRODUCTION

For a biologist intelligent behavior is behavior that is adaptive for the individual but also flexible and variable in a systematic and useful way in complex situations. Prerational intelligence is used here to designate mechanisms producing intelligent behavior without explicit use of symbolic thought and reason. Adaptive behavior or intelligent performance is evident at several levels of biological organization (e.g., at the level of single cells, individuals and societies), but the primary focus here is on individuals. The starting point is a list of attributes of intelligent behavior that is then used to compare intelligent behavior in animals, animats and systems from AI. The bulk of the chapter considers physiological and anatomical characteristics of animals relevant to intelligence, finishing with plasticity and learning. This review demonstrates the extent to which animals must be described as dynamical systems. Then different biological and artificial systems are used to consider levels of complexity and intelligence. The final section speculates about the relationship between prerational intelligence and the processes and mechanisms underlying cognition and logical reasoning.

In colloquial usage intelligence is intimately tied to humans and human behavior; it is closely linked to symbolic thought and reasoning. From the point of view of a biologist considering not just humans but the entire animal kingdom, intelligence corresponds to adaptive behavior first and to reason second. Thus, we will focus on intelligent behavior in the former sense and begin by identifying several attributes of intelligence. Adaptive behavior or intelligent performance is evident at several levels of biological organization – from large societies and small groups of animals through individual animals to subsystems within individuals – but we will focus primarily on individuals. Once we have a list of attributes, we can consider to what extent behaviors worthy of being designated intelligent can be produced without symbolic or rational thought. The term prerational intelligence is used to designate such behaviors and the underlying mechanisms. We will make an initial comparison of intelligence and system architectures as seen in classical AI systems, in animals and in animats. The bulk of the chapter will

11

J. Dean et al. (eds.), Prerational Intelligence: Interdisciplinary Perspectives on the Behavior of Natural and Artificial Systems, 11–87.
© 2000 *Kluwer Academic Publishers.*

consider characteristics of animals we feel to be basic or conducive to intelligent behavior. Intelligence in colloquial usage is intimately linked to
learning, so we will then discuss plasticity or modification through experience as it occurs in animals and outline some of the biological mechanisms.
This review suggests that animals must be analysed using the language of
dynamical systems.

Then we return to consider levels of complexity and intelligence in animals using artificial systems for illustrations. Finally, we conclude that, from
evolutionary and neurophysiological perspectives, we expect the processes
and mechanisms underlying high-level human intelligence, including cognition, language and logical reasoning, to be qualitatively similar to those of
prerational intelligence. Thus, many behaviors which humans naturally describe as logical processes in fact are realized in biological systems through
intelligence of design, as a result of evolution, rather than intelligence of
content.

2. INTELLIGENCE, INTELLIGENT BEHAVIOR AND RELATED ATTRIBUTES

A biological perspective on intelligence places more emphasis on function
and performance and less on a particular process, that is, the use of reason
and symbolic thought. McFarland (1993), in discussing rational behavior,
has referred to this distinction as intelligence of process and intelligence of
content, respectively, and we will use this terminology here.) As a starting
point, we characterize intelligence as adaptive behavior – that which is adaptive in the sense of furthering the reproduction of the individual and/or the
genes it carries, but which also carries the implication of adaptability, "the
ability to deal with new situations". The latter is the nominative use of intelligence (Lanz 1994); it clearly is an attribute of an autonomous agent which
is most obvious in action selection and modification (e.g., how rats behave
when confronted with new food items). In addition, there is the adjectival
use of intelligence: a statement, usually with an implicit value judgement,
about the characteristics of a particular behavior, action, design or process.
This latter use can be extended profitably to modules or subprocesses below the level of an autonomous agent, for example, to perceptual processes
organizing sensor data or to motor processes organizing movements. It can
also be extended to levels of organization above that of the individual, for
example, to groups or societies of individuals.

In the traditional approach to evaluating intelligence, the value judgement is provided by an observer and rests either on a notion of utility in a

function-oriented view, that is, a comparison of costs and benefits, or on a notion of efficiency in a process-oriented view, for example, intelligence in traditional AI as "the maximal application of available knowledge" (Newell 1990). Both judgements imply that the observer has rather complete information concerning the state of the agent and all aspects of the situation, something that does not apply to most natural situations. One biologically inspired alternative to this value judgement is to operationalize the evaluation in an evolutionary process whereby agents compete with one another and their actions affect their survival. In this view, one applied by Steels (1991) to autonomous robots, any adaptive behavior promoting survival is regarded as intelligent. However, this view ignores distinctive features in the common usage of intelligent. Thus, it precludes using these features to rank behaviors and examining the underlying control systems for structural and functional correlates of increasing intelligence.

For similar reasons, we adopt in this chapter a broad definition of cognition: following Newell, we consider cognition to involve the application of available knowledge. However, we include other forms of knowledge besides declarative knowledge. Examples are the top-down influences discussed in relation to recurrent nets and perceptual processes.

2.1 *Attributes Characterizing Intelligence*

The next step is to ask what features characterize intelligence or intelligent behavior in animals. In doing so, we put aside for the moment questions concerning the process or the origin of intelligence. The identification with adaptive behavior already emphasizes that intelligence is an attribute of an agent situated in a context. Our subjective judgements of intelligence are affected by aspects of both the agent and the context. The intelligence resides in the relationship between the set of behaviors that an agent could produce in a given context and the set it actually produces. Because we consider intelligence to be situated rather than an abstract generalized ability, we assume that intelligent behavior is only apparent in relation to goals or purposes, whether explicit or implicit, immediate or far-removed, and internally or externally specified. For animals, the goal is first and foremost increasing (inclusive) reproductive fitness. As an attempt to provide axes for the space of intelligence, we suggest the following attributes characterizing the agent, the context and their relationship. Some appear to be important in and of themselves; others become particularly important in competitive situations like natural selection.

Autonomy: An essential, if trivial characteristic, is that the agent perform on its own, utilizing external inputs as needed but without external guidance. More generally, one might insist that intelligent processes help maintain the autonomy of the agent or, when agents in a collective are under consideration, that of the larger unit (e.g., organism, society) of which they are a part. At the highest level, autonomy would involve identification, ordering and selection of goals and behaviors.

Complexity: Increasing complexity in the situation mastered is usually assumed to reflect increasing intelligence. Complexity (see Braun, this volume, for a formal discussion) can be expressed in terms of the number of relevant factors (goals, variables and value criteria or the size of the domain of possible actions) or the non-linearity of the relationship between inputs and adaptive outputs. Complexity in the task need not correlate with intelligence in the agent: because they are inflexible, hard-wired solutions (intelligence of process, either through design or evolution) are usually not seen as representing intelligence in the agent (intelligence of content, nominal intelligence). For the purpose of analyzing the organization of intelligent systems, however, such hard-wired solutions are equally interesting. Similarly, complexity in the agent's procedure is usually not regarded as intelligent if equivalent, simpler procedures are known; this point is related to the notion of efficiency.

Robustness: The behavior or performance should not be sensitive to random noise and variability in the inputs, outputs and intermediate elements. Neither should an intelligent system ever be at loss for an action. It should respond adequately to failure and not get trapped in endless cycles, that is, the system's behavior must be open-ended.

Adaptability and Flexibility: As outlined above, behaving appropriately in novel situations and making new responses in familiar situations are key features of intelligence. These features are related to the following three.

Generalization: Suitable generalization to new stimulus constellations is an important contributor to adaptive flexibility. The term generalization is used in two ways. It may refer to producing the same response to slightly different inputs, which can also be important in making the system robust. In a broader sense, generalization also refers to appropriate response modification, that which we include under adaptability. Applying a continuous transformation, be it a predefined function or an interpolation or extrapolation from previously performed input-output pairs, to new values in a continuous input space would represent very simple generalization (low intelligence). Bounded generalizations, restricted to appropriate input domains, would represent higher intelligence, which would be equivalent to category building. Generalizing into qualitatively new domains, if accompanied by appropriate controls and evaluations, would represent still higher intelligence and contribute to generating new behaviors and making the system open-ended. Insight is a high form of generalization.

Adaptive variability: Flexibility may also mean varying responses to identical inputs, which may make the behavior of the agent less predictable. This variability is necessary to provide alternative behaviors when generalization fails or reward functions change. However, it may also be important in exploration, as a way of maintaining optimal tactics and strategies and expanding the sphere of activity. Prior identification of alternative behaviors leading to the same end is a necessary basis for rapid, adaptive flexibility when conditions change. Whereas persistence in a highly efficient behavior may or may not represent intelligence, persistence in an ineffective behavior clearly indicates lack of intelligence and rapid switching from an ineffective to an effective behavior is generally a sign of intelligence. Stochastic variability would be a low-level variability, which may or may not be efficient. Learning and memory open a path to greater, directed flexibility.

Learning and adaptive plasticity over time: A particular sign of intelligence is the ability to adapt to changing conditions based on past experience. Parameter adjustment in a predefined space, as in self-calibration of sensors or effectors, would represent a low-level competence. Formation of new categories for perception, behavior or motor skills would represent a higher level. Reflexivity, permitting self-generated categories and behaviors themselves to become building blocks for new categories and behaviors, would be a still higher form. A prerequisite for learning is that the system have some failure tolerance. Some form of memory is also necessary to improve over stochastic mechanisms for avoiding persistence in ineffective behaviors.

Intentionality: By intentionality we mean directed, adaptive selectivity in the agent's procedures and in its environment. Examples would be identification of and selective attention to behaviorally relevant inputs, and selectivity in the choice of behavior—that is, of movement or more generally of perception-action patterns—based on internal states and external inputs. Brute force approaches, such as a stereotyped, exhaustive search of the space of potential actions, as opposed to selective search in regions where solutions are likely, would correspond to low intelligence. Active perception and exploration are examples of higher intelligence.

Anticipation: The ability of the agent to replace reliance on immediate inputs and feedback related to consequences of actions by predictive inputs and to replace action by simulated action in evaluating alternatives is also associated with increased intelligence. At a low-level, replacing reactive control by predictive control can lead to increases in speed and efficiency, but does not necessarily do so if accompanied by increased computational load. At a higher level, replacement of trial-and-error behavior by insight or appropriate generalizations is an analogous advance.

Utility: Because behavior is intelligent or not in relation to a purpose or function, intelligence can be evaluated according to the relationship between costs and benefits. For animals, this factor is particularly important in the allocation of time and resources to different activities.

Efficiency: For the same reason, behaviors that are faster or energetically more efficient for carrying out a particular task are generally considered more intelligent. Thus, faster algorithms or quicker problem solving are generally considered more intelligent, given equal levels of success.

2.2 *Comparison of Intelligence in Animals, Animats and AI*

Taken as a group, animals exhibit all of the characteristics of intelligence listed above, whereas animats and artificial intelligence systems still are limited in various respects. This is not to say that all animals exhibit all features of intelligence or that all technical systems show limitations in all respects. Obviously, many technical systems are highly successful and adaptive, showing many features associated with intelligence (see Ritter et al., this volume). The available space prevents detailed consideration of individual examples and all the possible exceptions, let alone an adequate discussion of the complete range of natural and artificial systems, so we will confine ourselves to features that contribute to the overall impression of a fundamental difference between animals and artificial systems. This initial comparison will focus on the level of individuals and complete systems, rather than on subsystems or collective behavior.

Most mature, non-social animals, of course, are autonomous by nature. They are not dependent on another animal for their supply of life-sustaining energy (energetic autonomy) or nutrients and they are not under the control of another animal (i.e., they have motivational autonomy, according to McFarland & Bösser 1993). (Social animals can be dependent on their social group to various degrees and, therefore, less autonomous.) For most animals most learning is unsupervised; there is no teacher. Autonomy also is a feature of ontogeny; biological systems are not constructed but must self-organize. This means that the intelligence provided by evolution is partly in the rules guiding development. In contrast, animats and artificial systems are by nature not fully autonomous in one or more ways. Reproduction, acquisition of energy, growth, development and repair are some features usually outside the sphere of action of most animats and artificial systems.

Animals and their behaviors are usually robust within the realm of naturally experienced situations. Robustness in control systems is provided by distribution of control functions over many elements. Loss of a few neurons may lead to degradation but not complete loss of performance. Often even considerable numbers of neurons can be eliminated with no measurable effect on performance. Robustness due to the presence of redundant control systems is also a characteristic of artificial systems where reliability is extremely important, as in airplanes and spacecraft. In such cases, simple duplication of function may be sufficient. In animals, for example in the leg coordination of insects, different mechanisms contributing toward the same end are often present.

Animal behavior is by nature open-ended and ongoing from birth to death. The nature of survival and reproduction means that changing the spatial or temporal situation may be sufficient to improve reproductive fitness. For many animals, locomotion or estivation provide ways to escape from unfavorable circumstances. From the point of view of technical systems, such a strategy might be equivalent to avoiding a problem rather than solving it. Because applied technical systems are oriented towards performing tasks under some kind of local temporal and spatial constraints, biological strategies optimizing a global function (e.g., reproductive fitness) may not be desirable or allowable. In other words, biological intelligence may be evident in the choice of problems to solve, whereas this typically is an external constraint for applied technical systems.

Complete autonomy, together with the ultimate goal of maximizing reproductive fitness, essentially combines all the situations experienced by an animal into one ongoing problem of enormous complexity. Thus, the task of survival and reproduction for most animals is a single, immensely complicated task, one involving many variables, opportunities and risks, alternative strategies and environmental situations that must be coordinated suitably within the budget of available time and energy. The optimization of time and energy expenditures is explored in behavioral ecology, but in other contexts these global constraints are often ignored and only the immediate situation is considered in evaluating an animal's behavior. However, even within a specific context, measuring all the factors affecting the behavior of an animal is difficult. The global optimization criterion means that intelligence in natural situations cannot simply be evaluated in terms of the immediate situation, because the situation for the animal is part of a continuum. Therefore, a complete evaluation of a particular behavior in a particular situation must take into account the effect on overall reproductive success. Most animats, in contrast, act in a situation that is significantly simpler because they are less autonomous and do not have to deal with many aspects faced by animals. Hence, the intelligence of animats can be evaluated more narrowly with respect to a given situation.

Flexibility in behavior is another feature of animals, but it is present to varying degrees depending on the niche and evolutionary history of a particular species. On the one hand, the ability to generalize from one stimulus to another and respond to new stimuli is universal; it is well documented in the literature on learning, ethology and experimental psychology. This literature also illustrates many of the underlying principles and the limits of generalization. Generalization is also a property of many artificial systems. In both cases, simple forms of generalization are a natural consequence of sensors that respond only to a subset of a continuous variable, for example, photoreceptors activated by a range of wavelengths. On the other hand, adaptive variability is not so universal. Exploration is a characteristic of many mammals and birds, but animals vary considerably in their ability to avoid persistence in an unsuccessful behavior. (Frequently such tests involve unnatural, restricted situations, so it is debatable whether anything can be concluded about the comparative intelligence of different species.)

Learning and plasticity are characteristic of virtually all animals. They also are characteristic of many technical systems. For both, learning and plasticity usually are restricted to particular domains. For animals, these are apt to be domains where the specific conditions to be faced by an individual cannot be predicted in advance or where specification of structure can be more exact by utilizing interactions with the environment in development. (Otherwise, genetic specification would be more reliable and incur less penalty.) For animats, the domains are selected by the designer. The difference in learning abilities between animals and animats is perhaps only one of scale, reflecting the more restricted problem faced by technical systems. Learning and plasticity are important to animals in many domains. These range from development through tuning of receptors and effectors to more complicated adaptations to a particular environment. As a first approximation, one can say that all activity in an animal leaves a trace and can change future activity. In contrast, man-made systems often have an artificial separation between execution and learning, with the latter embodied in special learning modules.

Anticipation is an important feature of animal behavior because it permits increases in speed and diminishes the risk of injury through faulty behavior. The use of internal circuits to replace peripheral feedback and the substitution of genetic specification for individual learning can be considered forms of anticipation. Associative learning is another major form of learning, which leads to the ability to anticipate. Again, artificial systems incorporate some of these features. AI systems rely on a world model to provide predictions. When the search algorithms are not efficient or applicable rules are not available, computational delays can be considerable.

Intentionality is apparent in exploratory behavior of animals and in active perception – both in the control of movable sense organs and in selective attention to incoming sensory inputs.

Utility and efficiency are criteria drawn from the evaluation of technical systems. Suffice it to say that careful studies of animal behavior in feeding and searching situations reveal an allotment of time and resources that fits predictions from standard optimization theories (Krebs & Davies 1978; McFarland & Houston 1981; McFarland 1985). Comparing efficiencies in time or energy is less useful for systems that are not in competition and use very different technologies. Competition among animals makes relative speed of performance particularly important. Technological systems are obviously faster in many respects and have displaced human or animal workers where they do compete, but often at the cost of increased energy usage.

The comparisons given above primarily relate to the behavior of individuals. However, most of the same criteria can be used to evaluate the performance of subsystems within an individual, like the network of neurons contributing to a particular process, or that of groups of individuals. In moving from one organizational level to another, an important question is whether new phenomena (emergent phenomena) arise over and above those conditioned by the increase in numbers of contributing units. Performance by groups of agents, such as ants constructing a nest, may be faster simply because there are more workers contributing. However, new phenomena may also emerge in changing levels; these may be advantageous or not. For example, the possibility of mutual obstruction or cooperation does not arise when only one agent is present. Whereas a single trail-laying ant is unable to optimize resource exploitation, the dynamics of the individual's behavior as part of a group helps optimize exploitation and trail length (see Section 7).

These qualitative comparisons indicate important differences between animals and artificial systems with respect to features related to intelligence, but of course one must remember that they compare systems arising under quite different conditions. Animals are the products of long evolution under conditions of competition and natural selection in a changeable environment. This constraint means that continuing persistence reflects an evaluation of the behavior and all other features of the animal with respect to their effect on (inclusive) reproductive fitness. The uniqueness of the niche of a species means that it is even difficult to compare intelligence among species, because each represents a long evolutionary adaptation to a particular habitat and life form, an important part of which is the interaction with numerous other evolving species. Most important, living in a real world demands that animals behave continuously; the time available for planning is highly variable, or conversely, the animal must be able to act in some way at any stage of the planning process. Although some artificial systems undergo competition and selection similar to biological evolution, this process usually is restricted to a subset of behaviors or single agents are evaluated on particular criteria. Adequate time usually is available for planning prior to action. Equally important, artificial systems or the conditions under which they evolve are by definition designed by someone, so it is easy to assume that any intelligence shown by the system is intelligence of design. The origin of intelligence in animals is not so easily penetrable. At present it is generally not possible to distinguish between intelligence of design and intelligence of content (McFarland 1993). (What is certain is that intelligence of content has arisen through one special kind of design process, that of evolution.) A final difference is that biological evolution imposes strong constraints on the types of changes that may occur. Each new organism must evolve from its predecessors via intermediate stages which themselves are successful adaptations. New features must be compatible with developmental constraints. Technical designs, on the other hand, can make radical breaks with previous artifacts and draw upon developments and technologies in wholly unrelated areas.

Besides these differences in origin, there are obvious differences in the hardware and in the underlying control schemes. The control schemes of classical AI systems rely heavily on symbolic representation and knowledge processing; the aim is to deduce intelligent actions by applying knowledge and rules. In many cases, inputs are already in symbolic form, so sensory processing is not needed. Where this is not the case, interpretation of sensor data is based on the same approach of applying rules to build from sensor signals to symbolic interpretations. Sensors and motors are assumed to have constant characteristics that are known to the controller. Metarules, rules about building rules, are necessary to make the system open-ended. Animal systems rely on muscles as effectors and neurons as control elements. Both the individual elements and the networks they form are plastic and modifiable in many different respects. Although these components are apparently inferior in some respects, the level of performance indicates that they have some advantages, which will be discussed in detail in the next section. Intelligence arises through patterns of activity in a complex network that may or may not encode rules in an explicit way. Animats and many control systems developed more recently in artificial intelligence adopt some of the features evident in animals, including a degree of self-organization, the use of approximate solutions and algorithms, and parallel and distributed architectures.

3. CHARACTERISTICS OF ANIMALS RELEVANT TO INTELLIGENT BEHAVIOR

We began with the remark that human beings are the measure of intelligence and showed in the next section that animals show many features of intelligence, particularly intelligence of process. Here, it seems appropriate to review some of the features of biological systems that appear relevant to this performance.

Consider a typical multicellular animal as an agent. Its capacity to act upon its surroundings is due primarily to the musculoskeletal system. Its capacity to be acted upon by the environment is due of course to the physical properties of the body but also to the presence of sense organs responding selectively to various aspects of the surroundings. The control of the output systems is the responsibility of the nervous and, to a much lesser extent, the hormonal system. Each of these systems has counterparts in technical systems. Nevertheless, it seems pertinent to examine the features of biological sensors, effectors and control systems, particularly because each component examined alone seems to be quite inferior to corresponding technical systems.

3.1 Component Properties

Muscles are – with a few exceptions (e.g., glands, intracellular pigment movements, cilia) – the primary effectors in most biological systems. Although muscles come in a wide variety of shapes and types, the constituent fibers are basically variations on a few themes (e.g., smooth versus striated, twitch versus non-twitch, fast versus slow). Speed of contraction and resistance to fatigue are the important functional variables: some muscle fibers can maintain moderate amounts of force for long times, while others produce large amounts of force for a short time before fatiguing.

Compared to motors, muscles seem to have many disadvantages. The force they produce depends in a complex way on the muscle length, the load, the speed of contraction, the exact pattern of impulses and the history of use. Past activity can lead to both short-term changes – like fatigue or potentiation – and to long-term changes involving growth and transitions between fast and slow types. Muscles must grow in size as an animal grows and they are subject to injury. These factors mean that muscle force and movement for a given motor command is not constant or simple for the control system to predict.

However, muscles do have several advantages over technical motors. The very fact that the individual fibers can grow and adapt their properties to the demands they experience provides a low-level self-tuning that simplifies the task of design and control. More important, muscles provide a highly compliant power system, so that unexpected obstructions do not immediately lead to damage and failure. The same spring-like characteristics lead to fairly predictable behavior under changing loads. Together with elastic elements in series and parallel with the muscles, the spring-like properties also provide an efficient way to store and recover mechanical energy.

Like muscles, neurons are very compact and complex elements with advantages and disadvantages. On the one hand, neurons are at a grave disadvantage in terms of conduction velocities and switching times compared to elements in electronic circuits. Moreover, most signals over long distances are transmitted using a pulse code of limited bandwidth, so only a few bits of information can be encoded reliably.

On the other hand, a single neuron is a compact processing element that can integrate a tremendous number of inputs in a sophisticated way (Llinás 1988; Shepherd 1990). Unlike typical artificial neurons, processing involves both digital signals for high-fidelity transmission of global results over long distances and extensive analog processing for distributed, local computations. In fact, complex morphologies can allow separate parts of one neuron to act effectively like different, loosely connected units. Moreover, neural processing can be modulated at several different sites. The following paragraphs briefly indicate some of the relevant features.

A typical neuron is morphologically specialized for both short and long-range interactions (Shepherd 1990). The short-range interactions are primarily mediated by the dendrites and soma. These elements were initially thought to be pure input elements, but it is now known that many also possess output synapses (Bullock 1979; Shepherd 1990). Cable characteristics, active membrane properties and cell morphology determine how electrical potentials spread, allowing inputs and outputs to affect and reflect, respectively, processing in the whole cell or in quite local neighborhoods. Synaptic output is a graded function of presynaptic potential, so these graded, local interactions represent a refined mechanism for distributed, analog processing. The general importance of these graded, analog interactions has become apparent only in the last twenty years. A variety of local interactions, including reciprocal synapses between two cells and synaptic complexes involving more than two cells are known (Shepherd 1990).

Long-range signals are transmitted over a specialized output channel, the axon. This activity reflects the sum of all the inputs to the dendrites and soma as they affect the potential at the base of the axon, which means the inputs are weighted according to the electrical characteristics of the neuron. Thus, the response at the axon to a particular input or pattern of inputs depends on the input type (excitation or inhibition), temporal pattern, location on the dendrites and soma and on the membrane characteristics of the dendrites and soma. Most axons, particularly those projecting over long distances, generate action potentials, which represent an all-or-none pulse code. As a result, the nature of the relationship between action potentials and information was the early focus of neurophysiologists (e.g., McCulloch & Pitts 1943). However, some neurons and receptors do not produce action potentials but rely wholly on graded potentials like the local interactions mentioned above (e.g., Roberts & Bush 1981).

Neural codes have several characteristics and limitations. One feature of the action potential, and some graded interactions, is non-linearity: action potentials occur only if inputs exceed a given threshold, a characteristic that, together with inhibitory influences, provides a natural way to gate activity in different pathways. Action potentials provide a fairly high-fidelity code for transmission over long distances, although there are examples where failure of transmission dependent on spike pattern is thought to be important. However, the low carrier frequency and the variability means that only a few bits of information can be reliably encoded by a single axon. Graded potentials suffer from similar limitations. Therefore, a more robust code is to link the message to the identity of the neuron (a labeled line) and use the level of activity as an indicator of intensity or certainty of the corresponding state. In this way, the message can be encoded by the particular pattern of active neurons within a large array and the resolution can be increased easily by adding neurons. Such a system is more robust; it is less affected by variation in the activity of single neurons.

Further modulation of transmission is possible at the output synapse itself by presynaptic influences that modify the amount of transmitter released by an action potential. This modulation of the output does not affect inputs and outputs at the soma and dendrites. Axons can project to several different targets and even send branches (recurrent collaterals) back to the neuron's own dendritic input region. The degree of divergence varies for different types of neurons; some project to discrete, localized targets whereas others

spread widely throughout the CNS (cf., fan-in/fan-out discussion in computer science perspectives).

Links between neurons, the synapses, also come in several varieties.
Neurons can be connected electrically through both rectifying and non-recti-
fying electrical synapses, but most synapses involve chemical transmitters.
Classical neurotransmitters interact with receptor proteins coupled directly
to ion channels (ionotrope receptors) to produce a short excitation or inhibition in the receiver by appropriate changes in membrane conductivity. Modulators, like serotonin or an ever increasing list of peptides, interact with receptors coupled to enzyme and second messenger systems (metabotrope receptors), which in turn affect ion channels to produce longer lasting changes.
Depending on the systems and channels to which a receptor is coupled, the
same transmitter can have quite different effects. Moreover, several varieties
of receptor molecules exist for the same transmitter. Synaptic transmission
is inherently non-linear, if only because effects on ionotrope receptors saturate as the postsynaptic potential approaches the equilibrium potential of the
moving ions and those on metabotrope receptors saturate when the target
enzyme systems reach maximum activity. Non-linearities due to thresholds
or interaction are also prevalent.

This description of ionotrope and metabotrope receptors already indicates that in the temporal domain a neuron, like a muscle fiber, is a dynamic
complex of interleaved processes related to its electrical, biochemical and
molecular states. Besides the Na and K channels typically responsible for
action potentials, neurons can express numerous other channels controlled
by membrane potential (voltage-gated channels), by classical transmitters
and modulators (ligand-gated channels controlled directly or via intracellular "second" messengers) or both. Appropriate channel combinations can
make the cell into a cellular oscillator generating rhythmic activity or cause it
to respond to a suprathreshold input with prolonged depolarization (plateau
potentials) and bursts of activity. Both types of channels can modulate intracellular messengers that in turn affect the state of cytoplasmic and membrane
proteins as well as protein synthesis and gene transcription. These intracellular mechanisms illustrate convergence and distributed processing. On the
one hand, several different excitatory neuromodulators can be coupled via
different receptors to the same second messenger system (e.g., Byrne et al.
1993). On the other hand, what from the outside appears to be a unified increase or decrease in response elicited by a transmitter can depend on several
different second messenger systems activated by different receptors.

Even this brief list of factors indicates that many different processes with
quite different dynamics contribute to a neuron's activity. Time scales range
from microseconds for conformational changes, milliseconds to seconds for

electrical potential and ionic changes, tens of milliseconds to minutes for biochemical changes and modulatory effects, to hours, days or a whole lifetime for structural rearrangements and growth and development.

The complex interplay of electrical, biochemical and molecular mechanisms means that activity or lack thereof leave traces in the that change its state, its shape and its response to subsequent inputs. Too much or too little activity in response to recent inputs usually leads to decreasing or increasing the responsiveness in a kind of self-tuning to keep the neuron's activity within a range where modulation is possible. However, too strong activation, as in a stroke, can also push the system into unstable regions where positive feedback leads to neuron death through over-excitation.

Like the other components, biological sensors have a similar combination of advantages and apparent disadvantages. On the one hand, they are extremely small compared to technical sensors and can be arranged compactly in dense arrays. Sensitivity can approach theoretical limits as in photoreceptors or chemoreceptors. On the other hand, specificity is usually relative rather than absolute. Furthermore, most receptors do not simply signal a single variable; instead they respond to some combination of the value of the stimulus and its rate of change. Adaptation, a sensitivity change occurring at the sensory structure, is usually present to keep the system within the working range defined by the limited band-width of the neural encoding mechanism. Because the action potentials only convey information about stimulus intensity, the quality conveyed by a particular afferent nerve fiber is not apparent in its pattern of activity but lies in the origin of the channel (labeled lines) or more generally in the pattern of activity across a population of such channels. The interpretation arises from the output connections the neuron makes. Variability in sensory responses, like that for motor systems, means that their interpretation by the CNS cannot be fixed; it must take into account previous signals and signals in other channels.

3.2 *System Properties*

Despite the complexity of a single neuron, the real power of biological networks arises from the assembly of many neurons into networks with inhibitory, excitatory and modulatory connections of different kinds and strengths. Examples are provided by so-called central pattern generators (CPGs) – single neurons (i.e., cellular oscillators) or circuits (i.e., network oscillators) that can produce patterned activity in the absence of structured inputs. Compared to cellular oscillators, network oscillators can produce more complex patterns with many different activity phases.

More generally, networks enable the massively parallel processing of inputs, outputs and internal signals. It is this characteristic of massively parallel structures that surely is one factor in the intelligence of animals. Certainly it contributes to the robustness of biological control networks.

Arrays of neurons, as already noted, also make it possible to encode parameter values with much better resolutions than is possible with a single neuron. A simple artificial example of such coarse coding (Baldi & Heiligenberg 1988; Heiligenberg 1994) is the use of the number of active units in a population of simple threshold units to encode the value of a parameter. Incorporating the graded activity of neurons in a weighted average further improves the performance. Maps, arrays of neurons maintaining topographic relationships present in the surrounding space or in the stimulus that determines their activity, are common in nervous systems. They range from the rough motor and sensory homunculi, the maps of the body present in the cortex, to topographic representations of external space or tonotopic ordering of acoustic units (Kandel et al. 1991). Coarse coding in such arrays permits hyperacuity performance, resolutions better than the size of the receptive fields of the individual sensors (e.g., Wehrhahn 1994). It is used in population codes in the motor cortex for specifying movement direction (Georgopoulus et al. 1993) and may be used in systems involved in orientation and navigation (Hartmann 1992; Hartmann & Wehner 1995). The latter theoretical work goes beyond mere representation in demonstrating how a mathematical operation, integration, can be implemented with arrays of neurons (see Section 6.4).

At the same time, massive parallelism in nervous systems does not imply a homogeneous structure. The division into anatomically and cytologically recognizable subunits – modules – is a prominent feature in the brains of higher animals. Modularity can be identified based on cell composition and form, patterns of connections within the region and to and from other regions, and even patterns of gene expression. The generic definition of a neural module is a population of neurons for which the interconnections within the group exceed those to neurons outside the group. Due to the prevalence of dendrodendritic connections, even a single, homogeneous layer of neurons might qualify as a module according to this quantitative criterion, but in practice modules are usually associated with circuits or heterogeneous cell groups thought to contribute to a particular process. Thus, modules usually refer to a cluster of elements identified on the basis of connectivity and implicated in a particular function. Obviously, the ratio of intra- and intermodular connections can vary in a graded manner, making modules more or less well-defined. Modularity often occurs on several different levels, particularly in the brains of vertebrates and those of higher molluscs and arthropods. In the primary visual cortex of primates, for instance, V1 receives thalamic projections related to a particular combination of ganglion cells, but within V1 there are columns and bands defined according to which eye is the source of the input, what stimulus orientation is most effective and so on.

Given the nature of neurons – with an axon for long-range communication and many dendrites for short-range, graded communication – and the way neurons arise in development, economy of wiring, particularly when many neurons are involved, favors grouping the neurons contributing to one process so the more numerous intramodule interactions can be performed using the simpler dendrites. Proximity also brings savings in conduction times that are not negligible relative to the speed required for some actions. A similar explanation probably applies to the common occurrence of topographically ordered projections, maps, in vertebrate and invertebrate nervous systems. Both in the sensory and motor systems, such ordering places closer together those neurons that are most likely to benefit from mutual interactions, be it for sharpening contrasts in sensory systems, coordinating activation of neighboring muscles or generally ordering the neural representation of the surroundings.

Given the prevalence of modules in nervous systems, the next question concerns their organization. Is it extremely hierarchical like some human organizations with a rigid separation of function and a strict chain of communication? Is there something like a general at the top, receiving condensed reports about the activity of far removed, low-level sensors and effectors and making strategic decisions while lower levels handle tactics and execution? One might expect the flow of activity to be primarily in one direction: beginning with the fine, local details registered by the sensors, proceeding to ever more abstractions at higher levels where patterns are stored and decisions about actions are made and then descending through various motor levels, becoming more specified until appropriate motor commands are developed. Both anatomical and physiological data indicate that such a hierarchical description is not accurate. Rather than a single chain of command, many chains are active in parallel but with extensive cross-connections. Furthermore, at almost every level activity passes not just in the direction from sensory to motor but also in the reverse direction. These pathways can be considered as feedback pathways informing the sender about the activity induced in the receiving module, but exactly what this reciprocal connectivity does is still unclear in most cases although a role in modifying and sharpening tuning has been demonstrated (e.g., Suga et al. 1997). In other words, each module, say in a sensory pathway, receives both bottom-up information from modules closer to the sensors in the periphery and top-down information from modules farther removed and concerned with more general features. Thus, the organization can better be described not as hierarchical, but as heterarchical (Turvey 1977; Minsky 1986) or hierarchical/parallel (Kolb & Whishaw 1990) with only a loose organization according to functional levels (e.g., Felleman & Van Essen 1991).

Similarly, strict separation along functional lines need not occur even within elements at the same structural level. For example, the ca. 400 neurons in each leech segmental ganglion produce appropriate movements for swimming, crawling, shortening and withdrawal (local bending) that require different patterns of activity in the muscles. Nevertheless, the interneurons connecting sensory and motor neurons are not strictly separated into groups contributing only to a single behavior; instead, many neurons participate in two or more behaviors (e.g., Lockery & Kristan 1990; Kristan 1995). In the stomatogastric ganglion controlling ingestion in lobsters modulatory transmitters can shift the activity of individual neurons, the ventricular dilator neurons, from a functional module associated with one behavior, the pyloric rhythm generator, to another, the cardiac sac rhythm generator (Hooper & Moulins 1990).

Modularity both in function and in anatomy also occurs within control systems for single behaviors, which raises the question of how separate modules are coordinated. An example is walking in insects (Wendler 1985; Dean 1989; Cruse 1990), where the circuits controlling the step movement of each leg are apparently localized in the segmental hemiganglion. Thus, there are six modules or step pattern generators, one for each leg. Experimental evidence indicates that there is no superordinate module responsible for coordination. Instead, the evidence suggests that appropriate interleg step timing arises from multiple coordinating influences passing through the connectives between the segmental ganglia. Furthermore, these interleg influences passed through the nervous system via the connectives are augmented by local responses of each module to the loading and movement of its leg which, due to the mechanical coupling through the substrate in normal walking, reflect the actions of the other legs. Thus, redundancy is present both in the centrally mediated coordinating mechanisms and in the presence of additional local mechanisms having similar effects.

Two temporal characteristics of neural activity are also important. Processing is usually asynchronous in that each neuron does its own integration and generates action potentials accordingly, irrespective of the activity of unconnected neurons. There is no central clock. However, this general asynchrony frees the temporal domain for potential use for other purposes. Intrinsic neural properties and connectivity can facilitate rhythmic activity (e.g., Llinás 1990). To what extent the fine structure of temporal patterning is important is still uncertain. However, one suggestion is as a marker of activity that is related in some sense, as described below (for a review, see Singer 1993).

Different areas of mammalian visual systems process different aspects (features) of a complex stimulus, such as its color, shape, texture, motion and location in the visual field. Typically, the adequate stimulus (i.e., the most effective natural stimulus) for a neuron in primary visual areas, which are only a few synapses removed from the receptors, is a quite simple stimulus (e.g., bars, edges) in a restricted area of the visual field (the receptive field of the cell). Cells in higher visual areas are specific for more complex stimuli defined by combinations of some simpler features (e.g., letters or even faces) and less specific for other features (e.g., location or color). This invariance can be obtained if outputs from many simple feature detectors with different receptive fields converge onto the same higher level cells. A natural visual scene presents a panorama of objects with different locations and combinations of features. Thus, the nervous system is faced with the problem of keeping track of what activity in different visual centers belongs to the same object in the external scene. One possibility is an explicit binding based on a direct extension of the low-level processing. This mechanism requires an additional module with cells that receive combinations of inputs from subsets of feature detectors and respond only when the specific combination is present (the pontifical cells considered and rejected by Sherrington 1941; the groups of cardinal cells favored by Barlow 1972; or the "grandmother cells" of colloquial usage). Activity in the high-level cell would signal the presence of the compound stimulus and could then initiate appropriate responses. Neurons apparently specific for faces have been described (Perrett et al. 1987; Rolls 1991), but they appear to be a special case. More generally, this mechanism would face a huge combinatorial explosion in the number of neurons required as the number of features to be integrated increases.

A more elegant solution is suggested by the synchrony observed in the action potentials of populations of neurons excited by a common object in the visual field. This synchrony, which makes use of the temporal domain of the action potential code, could solve the problem of binding together features processed in different areas (Eckhorn et al. 1988; Singer 1993). For example, the two halves of the visual field in cats and primates project to cortical areas on different sides of the brain. Obviously, it has significance for the animal whether a visual stimulus spanning the boundary between the two fields is a single object, which can be grasped as one, or two separate objects. It appears that interhemispherical connections in this case, and intercortical connections in general, enable a synchrony of action potentials (Engel et al.

1991). Environmental regularities again provide an important basis for the mechanism: physical objects exhibit coherent motion and usually are contiguous. The ease with which visual perception detects patterns of movement that are consistent with a single moving body (Johansson 1973; Ullman 1979) demonstrates this adaptation to the specific environment. Much evidence favors this hypothesis for feature binding, but it is not universally accepted and the extent to which it is important is still unclear. However, it seems plausible that synchronized activity in populations of neurons related to a common object would have a stronger effect on target networks, such as those concerned with directing actions involving the object.

Physical properties are also significant for the function of the motor control system. Animals are very good at utilizing the physics of their bodies and the musculoskeletal system to facilitate movement (McMahon 1984). Examples are the use of gravity to swing a limb forward in walking, the storage of kinetic energy in elastic elements for recovery at a later part of the step cycle and the use of mechanical resonance in the flight system of flies (Pringle 1957; McMahon 1984).

Control algorithms are also adapted to the considerable compliance provided by muscles and tendons. When collisions can be absorbed without injury, control algorithms can use approximations and let the compliance handle the resulting errors. In addition, the control task itself may not require high precision. The stick insect, for example, exhibits an interleg coordination which guides the tarsus of a leg in swing towards a position where the tarsus of the adjacent forward leg is standing and thus the presence of a foothold is guaranteed. Most substrates are larger than the tarsus, so it usually suffices for the moving leg to land in the vicinity of the target. Physiological evidence and simulations indicate that the nervous system uses an approximation in finding the desired final joint angles for the moving leg rather than an exact calculation of the trigonometric solution (Dean 1990; Brunn & Dean 1994). Thus, in considering the control of movement, the control properties of the nervous system cannot be simply considered in isolation; the properties of the controlled system and its interaction with the environment are equally important.

In summary, this review of the characteristics of biological control elements related to intelligent behavior suggests three features that are significant. First, the individual neurons themselves are highly complex physiologically, biochemically and morphologically. Second, the types of connections and the networks they permit are equally complex; for larger nervous systems, this involves complex modularity on several levels, but the organization is heterarchical with many reciprocal and recurrent connections rather than strictly hierarchical. Heterogeneity is present in the neurons, in the network architectures and even in the nature of the coding (e.g., labeled lines and temporal features). Finally, adaptations to make use of the regularities and physical properties of the environment are widespread.

4. LEARNING AND PLASTICITY

The word 'plasticity' is often applied to changes in behavior, as in behavioral plasticity (e.g., Section 7), but in this section it will be restricted to changes in the underlying mechanisms, that is, to changes in the structure or function of the control system not directly caused by injury. Thus, plasticity in the nervous system can underlie changes in behavior due to development and learning. These two forms play two different but related roles. They are similar in that both adapt individual animals to their environments in ways that cannot be fully specified in the genes either for reasons of capacity or because the relevant features of the environment are unpredictable. For example, developmental plasticity in the mammalian visual system can produce more accurate connectivity by taking advantage of regularities in neural activity that reflect regularities in the environment (Hubel et al. 1977). It also can lead to a better match between feature detectors and environment and to more economical representations of topographic regularities. Such developmental plasticity may be evident only once in an individual's life (Kandel et al. 1991) or it may continue in mature animals (Merzenich & Kaas 1982). This self-tuning is a useful form of intelligence of design; simulations show that it can occur without feedback.

Learning, on the other hand, is particularly important for continual adjustment to changing characteristics of the individual's interaction with its environment. Learning can be defined very broadly as any change in the behavior of an individual resulting from interactions with its environment that is not related to development, injury or fatigue. Learning from experience, the adaptive modification of an individual's behavior, is related but not restricted to intelligence of content. Some forms of self-tuning like those of development continue through life. Learning from experience, particularly more complex forms, is usually linked to some form of evaluation of performance generated either internally or externally. This type of learning requires some form of feedback about the performance.

Simple forms of learning, such as habituation and sensitization, occur even in organisms as simple as bacteria. Complex forms of associative learning, including blocking and secondary conditioning, are already shown by animals like snails, which are not generally considered paradigms of intelligence.

There have been numerous attempts to classify types of learning (e.g., Houston 1991). The classical scheme is based on the stimulus constellation. The primary categories are as follows: 1) habituation, a decrease in the response when a stimulus – usually an innocuous one of weak or moderate intensity – is repeated, 2) sensitization, an increase in the response to a test stimulus following exposure to a second stimulus – usually a strong or painful one – that does not depend on pairing the two stimuli, and 3) associative learning, either in the form of classical conditioning, involving associations between stimuli that are related in time or import, or in the form of operant conditioning, involving associations between stimuli and actions depending on their consequences. Associative learning has been the subject of most attention and research, because it can create new behavior patterns, a feature emphasized most adamantly in behaviorism (see Bridgeman et al., this volume). For the same reason, most forms of learning considered in AI and artificial adaptive systems fall into this category. Because these artificial systems are less autonomous than animals, they can utilize quite novel, non-biological forms of learning; in particular, the designer can give the artifact access to privileged information, even to the exact, correct answer as is the case in supervised learning. Unlike humans and many artificial systems, animals usually must rely on unsupervised learning involving less specific evaluations of performance.

An underlying assumption in classical psychological studies was that learning is a monolithic phenomenon, based on common mechanisms and applicable to any situation. One problem with this view was the limited time horizon of associative learning in laboratory experiments with animals: associations were best formed between stimuli separated by no more than a few seconds. Another problem with the standard classification was that individual examples of learning sometimes contained elements of several categories (e.g., classical and operant conditioning), and that some forms of learning did not fit easily into any category.

With the rise of ethology and the comparative study of animals in their natural environments, it became increasingly apparent that learning is more heterogeneous: each species has different predispositions concerning the conditions under which learning can occur and the types of associations that can be made. Two representative examples are taste aversion learning – the association of stomach illness with foods which may have been eaten many hours previously (Revusky & Garcia 1970) – and imprinting – the rapid, one-trial learning occurring in several natural situations (e.g., an ewe learning the odor of its new-born lamb (Kendrick et al. 1992)). Thus, learning is anything but monolithic in its phenomenology; to what extent this is true of the underlying mechanisms is not yet clear.

As studies of learning were broadened and refined, other distinctions became apparent and led to alternative classifications. Focusing on humans rather than laboratory animals suggested a set of categories based on neurological and behavioral evidence (Tulving 1987; Kupfermann 1991). Here, one important distinction is between procedural and declarative learning, often described as knowing how and knowing that (Ryle 1949; see also Lanz & McFarland, this volume). Procedural knowledge or skills – typically motor patterns like riding a bicycle but also perceptual discriminations like those of visual hyperacuity tasks – are typically learned slowly over many, many repetitions. Quite apart from its use in learning new skills, this learning can be important in continuously calibrating the sensorimotor system to adjust for changes in sensors and effectors. Such self-tuning is important even in animals like locusts (Möhl 1989). Declarative knowledge in humans is that which is typically expressed in words. It can be learned rapidly, even in single trials, and applied readily in contexts quite distinct from that in which the knowledge was acquired. Like classical and operant conditioning, however, these two types appear clearly distinct when extremes are considered, but the borderline becomes fuzzy when examined closely and the two types often complement and augment one another. Nevertheless, neurological evidence indicates that the two forms are mediated by different systems in the brain. Still other types of learning that are dissociable on the basis of neurological and physiological data include sensory learning – the brief record of sensory experience lingering for seconds after the stimulus ends – and some special forms like memory for faces one has seen (e.g., Damasio 1989; Kolb & Whishaw 1990).

The widespread occurrence of learning also indicates that mechanisms to support it are widespread throughout the animal kingdom. The simplest prerequisite is that each interaction somehow leaves a trace in the nervous system, even if the trace is no more complicated than that a particular synapse or neuron has been active. More complex forms of learning involve associations of stimuli, so they require a mechanism for linking activity related to separate stimuli.

Very early it was shown that learning and memory involve processes with two time scales (for a review, see Davis & Squire 1984; Kolb & Whishaw 1990). Short-term memory can be disrupted by interrupting the electrical activity of the brain and has a limited capacity; long-term memory requires protein synthesis for its formation (but see Hammer & Menzel 1995) and has seemingly unlimited capacity. This combination of short-term and long-term processes is universal; it apparently is a useful and efficient way to track changes in natural, noisy environments.

In thinking about possible memory mechanisms modern researchers used the metaphor of computers (e.g., Newell & Simon 1972). Two aspects of the metaphor are important and possibly misleading. First, computers have a distinct separation between data storage and data processing. Second, computer memory is organized using a system of addresses that is arbitrary in the sense that the physical location of stored information is irrelevant. Related pieces of information can be stored far from one another and information in adjacent locations can be totally unrelated. This model seemed to fit with the mass action and equipotentiality hypothesis (Lashley 1950) of cerebral function. However, the computer metaphor does gloss over two problems. First, the addresses themselves must be stored and accessed somehow. Second, memory must be extremely precise and reliable, because if addresses are lost, the information is virtually irretrievable.

Improvements in physiological and biochemical methods, as well as new technical developments, led to another model, one in which data storage and data processing are unified. Study of simple animals that are more amenable to experimental investigation than higher mammals but still show simple forms of learning – so-called model systems – contributed significantly to this understanding. The withdrawal reflex of the sea hare, *Aplysia*, shows habituation, sensitization and simple forms of associative conditioning leading to modification of response amplitude. These changes are both short-term and long-term. They have been traced to changes in the efficacy of several synapses in the normal reflex pathways, namely, those from sensory neurons to motoneurons and interneurons (for reviews, see Byrne et al. 1993; Hawkins et al. 1993). Here, the memory is a change in the process; there is no separate storage. The same is true for other examples, like one-trial learning in odor imprinting (Brennan et al. 1990; Kendrick et al. 1992).

Several additional characteristics are possibly significant. In *Aplysia* the mechanisms for plasticity are not related to the behavioral changes in a simple, intuitive way. Habituation and sensitization adjust responsiveness down and up, respectively; they appear to be simply different directions on a common axis. However, the underlying mechanisms are only partly complementary. Habituation reflects decreased synaptic transmission (homosynaptic depression) due to a depletion of transmitter vesicles available for release, whereas sensitization reflects increased synaptic transmission at the same synapses (heterosynaptic facilitation, because the sensitizing stimulus results in a second transmitter being released onto the presynaptic terminals) primarily due to an increased release of available vesicles, although it also increases the supply of releasable vesicles. Given the different time scales of the release and vesicle transport processes, this difference still appears reasonable in functional terms.

In other respects, newer experimental data have replaced initial, single-factor hypotheses with more baroque models. For example, the supply of vesicles depends on a non-linear interaction of several factors. At least two systems of second messengers (small molecules acting as intracellular signals) control different aspects of sensitization including sensory neuron excitability, supply of releasable vesicles and probability of vesicle release (Kandel et al. 1991; Byrne et al. 1993).

In contrast to the complexity of the mechanisms underlying simple sensitization, the mechanism that links activity in two pathways for associative conditioning is a model of simplicity and elegance (Hawkins et al. 1983; Walters & Byrne 1983; Yovell & Abrams 1992). The unconditioned stimulus is a sensitizing stimulus, but it increases synaptic strength more strongly in sensory neurons that were active shortly before its occurrence. Until complete removal of the calcium ions that enter a neuron during an action potential, this residual calcium marks the cell as having been active and amplifies the response to the nonspecific, widely broadcast sensitization signal (activity dependent facilitation: Hawkins et al. 1983, or activity dependent neuromodulation: Walters & Byrne 1983). The associative mechanism is in the amplified increase in the activity of the enzyme generating one second messenger; this increase is caused by the interaction at the enzyme of calcium ions, linked to a carrier, and the transmitter-receptor complex.

For the present discussion, details are less important than several general features. First, the combination of data storage and processing leads to non-arbitrary locations for stored memories, which in turn may facilitate information retrieval by content (as opposed to by address), a process more in accord with the subjective experience in attempting to recall a lost fact. Moreover, because learning is not a separate module but an intrinsic part of many modules for different processes, the capacity for plasticity can be adapted to the conditions of each specific task. For example, the mechanisms of synaptic plasticity in *Aplysia* were only discovered when synapses participating in a behavior showing learning were studied. Randomly selected synapses do not show such extreme plasticity.

Second, it is known that one of the second messenger systems, the cAMP system, is a common target of several different influences that increase synaptic transmission (Belardetti & Siegelbaum 1988; Byrne et al. 1993). Thus, different neurotransmitters and hormones interacting with different receptor molecules in the membrane produce the same effect by modulating the same intracellular pathway. Besides economy, one advantage of this convergence might be that saturation in the enzyme's activation places a natural ceiling on activation, avoiding the over-excitation that might occur if multiple influences combine additively.

Third, levels of responsiveness are often set not simply by the level of a single control variable but by two opposing variables, one acting to increase and one acting to decrease responsiveness. These two influences may or may not act additively.

Fourth, in a similar way, the behavioral effects reflect not one factor but the sum of several different changes, adding redundancy in the system (Goldsmith & Abrams 1991; Ghirardi et al. 1992). Increases in the behavioral response in *Aplysia* are based on the sum of several different effects: not only increases in the amount of transmitter released but also in the excitability of the sensory cells themselves. In a distributed network, many of the changes accompanying learning may be so small that they are below the resolution of physiological methods (Lockery & Sejnowski 1993).

Fifth, the same second messengers that mediate the short-term effects by changing the state of membrane and cytoplasmic proteins also convey the signal to the cell nucleus in order to modulate gene transcription and protein synthesis, leading to long-term changes in the levels of receptors and other proteins and eventually to changes in morphology through synaptic and neural growth. However, because concentrations of second messengers in different parts of the cell reflect the balance of synthesis, removal and diffusion, these short-term and long-term influences show different time courses and different dependencies on stimulus duration and repetition rates. One of the long-term changes is an increase in the activity of the key enzyme mediating the effects of one second messenger. Curiously, this increase is achieved not by increasing the concentration of the enzyme itself but by decreasing the level of the inhibitory regulatory unit (Bergold et al. 1990). The increase in enzyme activity, with associated effects on neuron activity, is produced first by transient increases in the second messenger concentration and then, if stimulus conditions are sufficient, by long-term decreases in the level of the enzyme's regulatory unit. Thus, short-term and long-term mechanisms are not completely distinct; the biochemical chain processing the incoming signal and the stored modifications share many elements.

Sixth, like development, long-term learning involves morphological changes in neurons and synapses. In both cases, matching growth or regression in the presynaptic element to changes in the postsynaptic element requires a flow of information in the direction opposite to the synaptic transmitter. In other words, retrograde signals must be present.

Finally, the biochemical processes supporting learning are not exotic; learning does not rely on special proteins or storage molecules. The relevant second messengers are integral parts of many regulatory pathways not just in neurons but in other cells. Thus, it is not surprising that learning is so ubiquitous, given that it can use such ubiquitous cell machinery.

This is not to say that all animals showing learning possess identical learning-related mechanisms. In particular, many vertebrate neurons show a phenomenon called long-term potentiation that, unlike the *Aplysia* example, appears to use a postsynaptic mechanism to specifically strengthen input synapses active when the postsynaptic neuron itself is strongly active. This mechanism is based on a particular type of glutamate receptor that has not been found in invertebrates. It is very much like the mechanism postulated by Hebb (1949) for associative learning. In contrast, the activity dependent presynaptic facilitation in *Aplysia* depends on activity in the presynaptic cell, not in the postsynaptic cell; it does not fit the Hebb scheme (Carew et al. 1984). Thus, learning mechanisms may be heterogeneous across different species and different systems.

Considering both learning and developmental processes together reveals another axis of heterogeneity. Both processes utilize regularities in the environment, which means that the mechanisms involve detection of correlations in time and space. However, correlations occur on different scales and between different modalities. Some correlations are quite local and involve signals of similar quality, as in the activity of two thalamic neurons projecting to the primary visual cortex and possessing adjacent or overlapping receptive fields. Both signals are tied to a rather specific context. In such cases, simple correlation of activity alone may be sufficient to elicit plasticity and lead to the formation of ordered maps. Other correlations involve signals with quite different qualities or scales. For animals, the most global scale concerns signals related to the fitness of the animal. In the *Aplysia* example, the correlation is between sensory activity in a particular pathway signaling an innocuous stimulus and subsequent activity in another pathway signaling a noxious stimulus that results in a widely broadcast, non-specific message about the state of the animal. In threatening situations, a host of hormonal and neural signals related to preparation for fight or flight responses are available to perform a similar role. Besides being associated with objective properties like modality (visual, acoustic, etc.) and intensity, many stimuli are associated with a subjective parameter ("pleasure" Cabanac 1991; "utility" McFarland & Bösser 1993; or the "reward" in reinforcement learning systems) that evaluates the referent according to an innate or learned scale. All these signals involve a kind of generalized evaluation of the state of the animal and its potential or actual change quite apart from a specific context. Because the evaluation signal is so widespread, a learning mechanism distributed in many cells can detect correlations related to global patterns of activity. This appears particularly important for learning in a highly modular system, where activity in many different systems must be linked with an overall evaluation of the animal's state and its change.

The impression to be gained from this brief survey is that activity in neurons marks them in many different ways. These traces occur on different levels ranging from concentrations of ions and second messengers, enzyme activity, and protein configurations and state to protein synthesis and neuron morphology; they have different time scales and different dependencies on the temporal pattern of activity (Byrne et al. 1993). Thus, neurons that evolution has imbued with plasticity respond in a continuous, dynamic way to the inputs they receive and to the outputs they produce. There is no discrete decision to store a particular pattern of activity; its very occurrence leads to its being stored to a greater or lesser extent.

What implications do these findings have for the nature of intelligence in animals and for possible applications to machines? Certainly the learning performance of animals is sufficient to attribute to them intelligence of process, where evolution was the designer. Nevertheless, from the standpoint of a human designer, the complexity of the mechanisms appears anything but simple, elegant and easily understandable, that is, well-designed. Instead, it appears that opportunism played a large part by using any elements that were readily available and contributed even ever so slightly to improving function. On the one hand, this does raise the possibility that just as behavior subsumes the action of many different mechanisms, these mechanisms themselves could be replaced by one or a few mechanisms generating the same behavior. On the other hand, the multiplicity of contributing mechanisms may be necessary to achieve successful behavior. The multiplicity does suggest that many different time constants and parameters can be adjusted in the course of evolution. The fact that evolution has settled on a combination of long-term and short-term processes together with heterogeneity of learning mechanisms and specificity according to task ought to provide a qualitative guideline for designers of intelligent machines.

5. ANIMALS AS DYNAMICAL SYSTEMS

Previous sections have characterized intelligence not as an abstract quality but as one intimately connected to the interaction of animals with their environment. One traditional view of this interaction is that animals are information processors, using their sensors "to pick up information present in their surroundings" (see Lanz & McFarland, this volume) and processing this information in order to select actions. However, this formulation glosses over the fact that the types of information available to an animal are limited by the types of sensors it possesses. The available data may be further reduced by developmental and learning processes. More important in the present context, to speak of information is already to adopt a language colored by rationalistic terminology, because we think of information first in terms of categories, which may be represented by symbols or by signal levels or simply by the presence or absence of a signal, and we think of information processing as performing operations on these categories. However, in many situations (e.g., latent learning), it is difficult to identify or define the relevant information. Moreover, information processing connotes a fixed procedure that must be completed before an answer is available, as is the case for computer programs or the Test-Operate-Test-Exit (TOTE) procedure of Miller et al. (1960). The metaphor does not explicitly address the question of time, although formal treatments of computability do address this issue (see computer science perspectives).

A more neutral standpoint is to consider an animal as a dynamic system embedded in a continuous interaction with its environment that is governed by the animal's sensors and effectors and by the laws of physics. The animal's actions – its observable behavior – are dependent on both the inputs it receives and all the properties of the animal including its nervous system and the physical properties of its body. Thus, the animal is a system with a state corresponding to the physical state and to the current pattern of activity in the nervous system. Changes in state can be specified as a function of the state and the inputs, meeting the requirement for a dynamic system, but the function is complex, non-linear and usually unknown.

In principle, the state of the nervous system could be defined at several different levels. These range from the tendency of the animal to behave in certain ways as measured with behavioral methods (e.g., motivation) through the pattern of neural activity measured with physiological methods to details of the biochemistry and structure of the individual neurons that can only be measured indirectly in an intact, behaving animal. All of these levels are interrelated. Llinás (1988) has suggested that the state of the nervous system, in this sense of dynamic systems, may be the basis of subjectivity – the difference in responses to different repetitions of an apparently identical stimulus. Obviously, the more detailed the level of description, the more variables are involved, which increases the dimension or number of degrees of freedom. However, even for a simple animal a complete characterization of its physiological state is quite beyond current experimental methods, even with new optical and physiological techniques for recording the activities in populations of neurons.

The important fact from the discussion of the previous two sections is that nervous systems of complex animals process information in a distributed manner and that activity at one time affects the activity at later times in numerous ways. Even clearly measurable changes in behavior can result from small changes – many of which probably are not experimentally resolvable –

in the activity of many different neurons (e.g., Lockery & Sejnowski 1993). The accompanying changes on different levels in the nervous system occur with different time scales.

The role of time is particularly significant because a finite time is required for activity patterns in the nervous system to shift from one behavioral output to another. According to some estimates, at least 100 ms are required following the presentation of a visual stimulus before the resulting activity reaches the state at which recognition occurs, which suggests there are about 10 synapses or processing steps between sensory input and recognition (Oram & Perret 1992). Similarly, the readiness potential and associated changes in single cell discharges show that, several hundred milliseconds before motor output commands are actually sent, activity in motor centers starts to change to specify the form of the intended action.

In behavioral studies, humans can be forced to act before this specification is completed. These experiments (Favilla et al. 1989; Ghez et al. 1989, 1990) show that a simple movement command, in this case one for a force impulse with different amplitudes in one of two directions, evolves over several hundred milliseconds by increasing differentiation of a default action. Thus, this biological system does not need to wait until specification is complete; it can produce some action at any time.

This behavior of the motor system corresponds to the performance of the sensory systems when stimulus presentation is shortened. Thus, in both sensory and motor systems, a finite time is required to shift patterns of activity in order to achieve a complete, accurate representation of the stimulus or motor act, but partially specified responses of increasing accuracy can be made sooner. This is consistent with the selective pressures under which animals have evolved. Response speed is often important, so an animal cannot afford to wait until stimuli have been accurately identified and actions completely planned. Differences in response latencies and complexity do not just reflect the hierarchy from short reflexes in the spinal cord to complex planning in the cortex. They also reflect the time required for the cortex and other relevant centers to shift to an appropriate pattern of activity.

The experiments of Ghez and colleagues also show that the specification of the different response parameters, direction and amplitude, evolve separately (Favilla et al. 1989). This dissociation is further evidence for the heterarchical organization of the nervous system. Dissociations in the specification of different components of a movement or in the use of stimuli for movement control as opposed to conscious perceptual tasks reflect the concurrent parallel processing in different modules connected primarily to one or another output system. One of these outputs is the verbal reporter who provides a more or less coherent and rational summary when probed for a verbal description, the function of which will be considered later.

The dynamic systems approach provides a systematic formalism for describing the behavior of complex systems with many variables and processes. Its appeal is that it provides a descriptive tool that can be applied equally to all levels from neuron to neural modules to individual animals to groups of animals, as long as adequate describing variables can be identified. One example is the synchronized activity established in parts of the visual system processing a common stimulus. Modeling these networks as a dynamic system shows that the synchronization reflects properties of both the stimulus and the cortical connections (Lytton & Sejnowski 1991). A second example concerns the interaction of different modules on the same functional level. For example, the interaction of the step pattern generators producing the rhythmic stepping in each leg of the stick insect can be described as exchanging information (Cruse 1990; Dean 1991), but this description implies something is already known about the function of the influence. A more accurate description must encompass the dynamics of the signal's effect on the receiver. Similarly, when different modules compete for the same effectors, as when an animal selects whether to rest or look for food or which of several different visual stimuli to fixate, circuits within the nervous system must resolve the conflict. This process can involve summation, as when a fish combines gravitational and light stimuli in adopting a compromise posture (v. Holst 1950; Mittelstaedt 1964). It can involve mutual exclusion, as when a fly alternates between fixation of one or the other of two black stripes in its environment. A similar analysis applies to collections of ants adapting their trails to changes in the available food supply (Section 7).

Within this framework, one can also envision modules within the brain that perturb the activity of other modules, and thereby facilitate and amplify particular subpatterns, as in shifts of attention (e.g., Olshausen et al. 1993), or modify the thresholds and timing for initiating a behavioral output, as in the experiments mentioned above.

As an analytical tool, the theory of dynamic systems provides ways to quantify both the stability of a particular pattern of activity or behavior in response to internal and external perturbations and the time scale on which changes from one state to another occur. When experimental data are sufficient, the theory can provide predictions about the relationships of these different system properties to each other, for example, that long transition times correspond to high stability. However, while adopting this approach as an ideal, one must also admit that it can be exceedingly difficult to rigorously apply to complex systems like those involved in intelligent behavior.

6. LEVELS OF INTELLIGENCE: CORRELATIONS OF PER-
FORMANCE AND STRUCTURE IN IDEAL NEURAL NETS

Considering the spectrum of animals, one finds a broad range of complexity in both behavior and nervous systems. Early animal psychologists attempted to use learning ability to compare intelligence in different species (see Bridgeman et al., this volume). The theoretical basis was that learning represented a single universal mechanism that was applicable to any task and possessed by different animals to different degrees (a view discussed and criticized in Seligman 1970). The rise of ethology made clear that behavior, like morphology, represents a specific adaptation to a species' own particular niche. Strategies of different species form a continuum from specialists adapted to a narrow range of habitats to generalists able to survive under many different conditions. Learning, as described in Section 4, was no longer seen to be monolithic. Instead, each species has different capacities for plasticity and learning; each has different predispositions for responses in new situations and for learning different tasks. Thus, one cannot compare species on an abstract scale of intelligence (Hodos 1982).

Nevertheless, species differ in the extent to which their behavior shows the characteristics of intelligence discussed in Section 2. In looking for organizational differences correlating with these behavioral differences, one might begin with comparisons of primitive and advanced animals. However, all existing species are the product of equally long evolution, albeit under different conditions. Thus, in a strict sense, it is not correct to speak of primitive and advanced animals. Primitive means only that we believe, usually on the basis of morphological or genetic evidence, that a particular species is closer to what common ancestral forms are thought to be like. Moreover, evolution does not proceed with continuous, equal advances in all attributes. Different evolutionary histories emphasize flexibility in some domains and not in others. Thus, a digger wasp may appear very intelligent in memorizing the location of its burrow and very stupid in repeating the investigation of its burrow simply because the prey left at the entrance during the inspection has been moved slightly by a scientist (Fabré 1989).

For these reasons, existent species do not form a clear progression of increasing intelligence. Therefore, it is easier to consider hypothetical stages in intelligence and possible organizational correlates on the basis of artificial systems and to augment this discussion with biological examples where appropriate. Different architectures for animats and artificial neural networks provide clear examples, whereas actual biological networks are usually heterogeneous and certainly less well understood. Animats are appropriate examples because many are constructed to simulate essential properties of real animals. Artificial neural networks are appropriate examples first because animats are usually controlled by artificial neural networks and second because artificial neural networks are simplified models of biological neural networks.

Thus, we can consider a hypothetical series of animats showing an increasing number of features and abilities that make these animats more and more similar to the higher animals and humans that we believe to be intelligent.

6.1 Feedforward Networks and Reactive Agents

The simplest animat is a reactive agent in which the effectors are driven solely and directly by the sensors. The corresponding neural network, biological or artificial, is a strictly feedforward net coupling inputs to outputs. In theory, such feedforward nets can approximate any continuous input-output transformation. The goodness of the approximation, of course, depends on the size of the net and on the presence of an appropriate means to structure it. However, even nets of two or three layers structured with standard algorithms are sufficient for good approximations to transformations needed in controlling movement and in discriminating among patterns (see Braun, this volume).

The archetypes for these two tasks, approximation of a continuous function and classification, are two ends of a spectrum representative of all natural tasks. Significantly, improvement in the approximation for a given architecture can be obtained by using output units with activation functions (linear, logarithmic, heaviside) corresponding to the type of task.

Many sensorimotor transformations are smooth and continuous in the sense that small changes in the input correspond to small changes in the output, so even a linear approximation is good locally. For each arm position, for example, small changes in shoulder and elbow angles produce small changes of predictable directions in the position of the hand and these changes combine vectorially. Feedforward nets can be structured to reproduce these transformations and used to control robot arms.

Classification, on the other hand, requires abrupt boundaries between domains. Feedforward networks with or without hidden layers can approximate classification; that is, they can produce discrete output patterns that divide the input space into discrete domains. The classification can appear explicitly only at the output, in which case the network can simulate "choices" among different behavioral responses. In networks with hidden layers the classification can also become explicit at a hidden layer and then be used either to determine behavioral outputs or as inputs to subsequent hidden units. In sensory processing, for example, complex feature detectors could be implemented in units receiving their inputs from elementary feature detectors in prior layers, units that respond to individual stimulus components. This architecture could provide one solution to the problem of feature binding, that is, the question of how different aspects of a complex stimulus are tagged as belonging together (Section 3.2.). Activity in a unit combining inputs from units responding to blue, to circles, to contrasts at a particular location in the visual field and motion in a common direction would explicitly represent the presence of such a compound stimulus at a particular location. Often this architecture is designated using a hypothetical instance: "grandmother cells", individual cells responding only to the combination of features shown by a particular grandmother. However, this solution suffers from a combinatorial explosion in the required number of units as the number of features and combinations increases; it is also brittle unless redundancy is added by including multiple copies of each highly specialized cell. Of course, specialist grandmother cells could be replaced by patterns of activity in the target layer as an intermediate, distributed representation. This coarse coding would enhance robustness and generalization, but it faces problems of storage capacity. Biological networks may take advantage of the temporal structure of neural activity to achieve a more elegant solution – one avoiding the need for an intermediate layer where features are explicitly combined – to this problem of tagging related features (the feature binding problem discussed above in Section 3.2).

Several specializations, both on the sensory and motor sides, can improve the performance of feedforward nets. On the input side, the introduction of spatial high-pass filtering (lateral inhibition) can make the system independent of absolute stimulus intensity and enhance the detection of borders. Simple filters can perform tasks superficially requiring logic. An example is the distinction between self-induced motion and global field motion in flies (Egelhaaf 1987): fast global motion of the visual field, which usually corresponds to an abrupt, active turn by the animal, elicits larger responses in one type of cells, whereas slow motion of the global visual field, which usually results from motion of the surroundings, elicits larger responses in a second type of cell (Egelhaaf 1990). The latter stimuli elicit compensatory turning (optomotor responses) to stabilize the visual environment; the former do not. Using appropriate training techniques (see below), feedforward nets can also be structured to take into account relationships inherent in the topography or temporal order of incoming stimuli. The potential advantage is a simpler structure for the network or a reduction in the dimensionality. Topographically ordered projections in many sensory and motor systems and in Kohonen networks are examples.

In isolation, a simple feedforward network has no significant dynamics, because the output is determined by the input. Nevertheless, when such a network is placed in an agent that can move and act in an environment (i.e., a so-called "situated agent" forming a recurrent "loop through the environment"), the dynamics of the interaction can lead to complex behavior. This is because the output of the system affects its subsequent input, leading to a new output and so on. Depending on the reaction, this interaction between agent and environment can stabilize a particular relationship or produce long sequences of different actions. Extremely simple artificial examples of such reactive agents are the vehicles envisioned by Braitenberg (1984). Selection of different outputs can be performed simply by integrating all different influences at the effector, as when the output corresponds to speeds and directions for the motors on two sides of a machine (e.g., Webb 1994). Motoneurons perform a similar integrative role within a restricted domain, a fact reflected in Sherrington's description of them as the "final common path" for influences arising from different sources.

This reactive architecture corresponds to two possible uses of sensory input in animals. First, sensory feedback reflecting the consequences of an action can be used continuously for comparison with a desired value (called the set point if constant or the reference input if variable over time) in feedback loops for stabilizing particular relationships ("continuous use" or Type I, Cruse et al. 1990). The notion of feedback control or servocontrol is very important both in technical systems and in animals. Simple instances are resistance reflexes like the mammalian stretch reflex, where a change in muscle length due to load variations elicits a change in muscle activation that tends to counteract the length change. In a strictly feedforward net, the set point might be identified with the pattern of biases on the units. For a net where inputs representing different stimulus parameters converge onto common outputs, slowly changing inputs from one modality might be regarded as setting the bias for those of another, making the behavior of the feedforward system more variable and adaptable.

Second, sensory feedback can be used to trigger a new action ("advance use" or Type III, Cruse et al. 1990). Strictly feedforward connections occur in simple reflexes, escape responses and some taxes. The nervous system of coelenterates (Bullock & Horridge 1965), with few layers and some direct coupling of sensory to muscle fibers, is closest to this strictly feedforward net. This is sufficient for simple escape reflexes and taxes. Active and passive obstacle avoidance is also possible, as shown by Franceschini et al. (1992) in a vehicle using simple movement detectors modeled on the compound eyes of insects. Locomotion itself usually involves a further level of complexity discussed in the next section: either internal default actions or, in animals, intrinsic rhythm generators.

At the level of individuals and groups, feedback through the environment is the essence of stigmergy, the system controlling many features of the collective behavior of social insects (see Section 7). Like a Turing machine, the ants write messages that in turn are read and affect future actions.

Thus, even simple feedforward networks demonstrate some of the features we have identified with intelligent behavior. They are able to solve problems of great complexity. They generalize to new inputs by producing smooth interpolations between previously produced input-output pairs. Coarse coding of inputs and intermediate signals, using many units with low precision plus a procedure for interpolation, is an alternative to precise, analytical algorithms (e.g., Lehky & Sejnowski 1990). Generalization can also take advantage of redundancy in the input signals, so that the system is robust when sensor signals for one feature or another are missing or noisy. More generally, nets with more than a few units are robust; they can tolerate faults in a few components.

Reactivity, using feedback of motor output via the environment, can provide simple feedforward systems with flexibility and adaptability if the control system is organized in such a way that all different constellations of agent and environment are foreseen. An example is the step coordination of insects where loss of one or more legs often occurs through natural causes. Following leg loss, the coordination of the legs adopts a new pattern that ensures stable support under the new conditions. This is an example of adaptation that does not involve changes in the short-term or long-term structure of the controller in the central nervous system. Neither explicit recognition of the new situation, that is, recognition that a particular leg is missing, nor learning new strategies appears to be necessary; the original control rules are formulated sufficiently generally that the altered sensory inputs elicit appropriate leg coordination.

However, the output of simple feedforward networks is not flexible in the sense of being able to vary the output for a constant input. This means that when there are different output patterns consistent with the task or inputs, in other words, when there are redundant degrees of freedom, the feedforward net only provides the one that was designed or learned. Various techniques can be used to select the output that is best in some sense, so the network can show some intelligence of design. This optimization of design can specify inflexible, preselected motor programs for a given task or implement fixed principles, such as applying a cost function to optimize the performance according to one or more criteria. Nevertheless, the network cannot produce alternatives and provide the flexibility characteristic of intelligence. There also is no reflection on the output – no feedback or evaluation of its quality. (We are ignoring for the moment any training scheme used to establish the weights in the network.)

6.2 *Feedforward Networks with Intrinsic Activity*

Up to now only systems have been considered that are purely sensory driven. An important property of most nervous systems is to contribute intrinsic or autonomous activity generating behavior in the absence of sensory input; such activity is a property of even simple coelenterate nervous systems (Bullock & Horridge 1965). As indicated above, the interaction or loop through the environment can already be used to make a situated feedforward control system recurrent in a sense and allow it to stabilize particular states over time or produce sequences of outputs. This variability in the output is possible even though the weights in the net do not change.

An important advance in this control is to permit another instance to modify the desired value of the controlled parameter. An example that is particularly clear because the bias is set directly at the sensors and the comparison between actual and desired values occurs in the periphery is the stretch reflex mediated by the vertebrate stretch receptor. The gamma activity to the small muscle fibers in the spindle alters the response of the primary (Ia) sensory afferents and in turn their effect on the alpha motoneurons controlling the main muscle fibers. The gamma activity thus establishes a set point for activating the stretch reflex to oppose changes in muscle length. (Although in theory this circuit could provide complete servocontrol of muscle length, the gain is such that only a servo-assisted control is realized: as often occurs in biological systems, the responsibility for accurate control is distributed among several different mechanisms, including a fast feedforward pathway and several feedback pathways of which the stretch reflex is just one.) Unlike this mammalian stretch reflex, the set point or reference input in most resistance reflexes and other feedback circuits is specified and compared to the measured value within the central nervous system. Central specification provides a quicker way to vary the set point, and thus the equilibrium point of the system, in accord with the activity of other subsystems.

More complex temporal behavior can be obtained if the reference input changes continuously with time. If the output of a central pattern generator (CPG) is used to vary the bias and provide the reference input for comparison with the input signals in a feedforward net, complex temporal behavior can be obtained even in the absence of changing inputs. As mentioned above, a CPG can depend on properties of a single cell or properties of a network or a combination of the two. Such networks usually involve recurrent connections, such as mutual inhibition, but for the present purpose we want to ignore how the temporally varying reference input is generated and consider the CPG's role in the network.

In a sense, a CPG represents a match established through evolution between the output pattern it produces and the input-output relationship that the animal will actually require. This match can be considered to be an implicit model of part of the world, a model that supplies predictions about temporal and spatial regularities in the animal's interaction with its surroundings. Thus, it is not surprising that CPGs are particularly well-developed for motor behaviors performed under homogeneous conditions (e.g., flying, swimming, swallowing, breathing). For example, the air in which a flying locust moves is a fairly homogeneous medium, so the effect of activating muscles to elevate the wing almost always is that the wing will be elevated after a predictable delay by the specified amount. The advantage of a CPG is the reduction in time delays by omitting the loop through the environment via effectors and sensors. As a result, central pattern generators are almost universal for basic, rhythmic behaviors (Delcomyn 1980). The prediction also reduces the reliance on proper sensor function implicit in a purely reactive system. If the prediction is good, a CPG represents an intelligent control strategy, but alone it does not show the flexibility and plasticity characteristic of high intelligence. Usually, adaptive modification of the CPG itself is necessary, but this, too, can be achieved by feedforward connections or, as will be discussed below, by recurrent connections. In the locust example, the pattern of activity in central pattern generators is subject to modification by sensory inputs. If, for any reason, the wing of the locust is elevated too much or too soon, proprioceptive feedback can initiate a correction and also shift the phase of the central activity (reviewed in Wendler 1985).

Adding intrinsic activity gives the system "state", meaning that the behavior of the system is not determined solely by its inputs. Not only can the intrinsic activity from the CPG simply summate with the input activity to produce new patterns, it can also modify in a variable manner the gain of the connections from inputs to outputs. Thus, even in a strictly feedforward network, the CPG can gate input activity. The non-linear, threshold response function of neurons with action potentials provides a natural way to gate activity in different pathways, as has been shown in the locust flight system (Reichert & Rowell 1986). Gating may also be useful in implementing an intermittent feedback control.

Obviously, any feedback through the environment will make the dynamics of the interaction between central activity and inputs much more complex and require new techniques for analysis. Beer (e.g., Beer & Gallagher 1992; Gallagher & Beer 1993) discusses ways to analyze these dynamics. However, if the internal component (the CPG) is not affected by the input – not the usual case – such systems are still basically feedforward systems with no feedback control of the actual output; they cannot exhibit the reflexive, flexible behavior of intelligent systems, although their behavior may be highly complex. For example, such a feedforward system has no way to discriminate between input changes arising from external changes and those arising from the actions of the agent. (Whether this is important will be discussed below.)

6.3   *Recurrent Networks*

A network is recurrent when there is a pathway, excluding that via the environment, by which the activity in a neuron can affect its subsequent activity. This feature leads to complex dynamic behavior in recurrent networks. Recurrent connections are the basis of network oscillators and other CPGs; complex recurrent networks can produce arbitrarily complicated, even chaotic temporal patterns.

In a formal sense, many networks with recurrent connections can be reformulated as strictly feedforward networks performing the same input-output transformation. One technique is unfolding a recurrent network in time by considering its output at each time interval t as the input at time $t+1$ and training a feedforward net with this set of input-output pairs (Minsky & Papert 1969; Rumelhart et al. 1986). Similarly, lateral inhibition can be implemented either in the pattern of feedforward connections from one layer onto a target layer or in terms of recurrent connections among members of a single layer. Although these different formulations are formally equivalent when connection strengths (weights) are appropriately chosen, there may be differences in stability for non-optimal connections and formulations using recurrent connections can be significantly more compact.

Moreover, biological systems do not appear to use recurrent and feedforward nets interchangeably. In fact, recurrent connections are a dominant feature of the brains of higher mammals, where for instance most cortical areas receive a return connection from the areas to which they project (e.g., Van Essen et al. 1990; Felleman & Van Essen 1991) and long recurrent pathways involving several relay stations are also common. The motor loops between cortical motor areas and the cerebellum and those between the cerebellum and spinal motor areas are just two examples (Brooks 1986).

A significant feature of recurrent connections is that they can replace or augment the coupling through the environment described in the last section. By reversing the forward flow of activity from sensory to motor elements or by connecting two areas on the same level, they facilitate the establishment of activity patterns that are internally consistent within modules and mutually compatible between modules. Recurrent connections between higher and lower sensory centers permit higher levels, where global or invariant features are represented, to influence lower levels (top-down influence). In this way, activity in local feature detectors that is ambiguous or noisy can be coerced towards patterns consistent with the overall activity of the population of local units as reflected in the activity in the higher layer. Such a top-down influence is an example of a cognitive process because global information or hypotheses, represented in the connectivity of the higher levels, is used to bias or interpret the incoming pattern of activity on lower levels. Similar interactions surely play an important role in the interpretation of spoken or written language, where ambiguities at different levels ranging from acoustic to semantic to grammatical must be resolved in a way consistent with the overall pattern. Recurrent connections presumably are also active in switching attention among different parts of the sensory input in an internal equivalent of orienting moveable sense organs: both are features of active perception, which is one expression of intentionality.

A special case is recurrent activity arising from motor elements, which is called either a corollary discharge (Sperry 1950) or an efference copy (v. Holst & Mittelstaedt 1950). These signals provide other areas in the brain with a prediction of what the motor system is about to do. They can perform at least three functions (Bell 1984). According to the original definition, an efference copy represents a quantitative prediction of what the reafferent sensory input will be. By comparing sensory signals with a centrally produced "expected" value, the CNS can distinguish between relative motion caused by movement of the environment and that caused by movement of the agent. (By enabling this distinction, reafference can contribute to a sense of self.) Efference copies can be used to cancel out self-induced apparent motion in order to maintain a stable frame of reference. Corollary discharges, which we will use as the more general term, can also be used in a less quantitative way to heighten or depress the sensitivity of sense organs that will be affected by an impending action. All three functions are present in some weak electric fish, which use a pulsed electric discharge for communication and for sensing objects in their surroundings (Bell 1982). Corollary discharges linked to each pulse briefly increase the sensitivity of the nucleus that evaluates modifications in the field due to surrounding objects. They depress the sensitivity of the nucleus processing communication signals from other conspecific animals and thus preserve its sensitivity in the intervals between pulses. A signal to a third nucleus, which is involved in detecting weak external sources, quantitatively cancels the effects due to the animal's own pulse.

By analogy, other recurrent networks within the brain may perform similar functions. When the recurrent connections are collaterals of the primary output units, they will provide a copy of signals the network is sending to other areas. If they arise from other neurons within the module, interpretation is more uncertain. The recurrent signals might play the role of either sensory reafference, signaling the state of the module's activity, or that of the corollary discharge, preparing other regions for effects related to the module's primary output.

Networks with lateral connections will be considered as another special case, although in a formal sense and in artificial neural networks one can argue that lateral connections are actually a special case of recurrent connections that happen to return to the layer from which they arise. Nevertheless, in biological networks, it is often convenient to speak of lateral interactions between neurons and circuits on the same level in a hierarchy of sensorimotor processing.

Such lateral interactions, whether reciprocal or not, often contribute to forming network CPGs using cells that are equally far removed from the final motor output. More generally, when modules take the place of neurons, lateral interactions can be used to select among alternative, mutually incompatible states of the system. Networks proposed by Maes (1991) and others use inhibitory and excitatory connections for making behavioral decisions. By having modules – command neurons or command networks – that interact with each other in suitable ways, the state of the whole system can be shifted appropriately to prevent simultaneous execution of behaviors that require different use of shared effectors. Such interactions also can be used to prepare ("prime") sequences of activity that correspond to the most likely sequences of motor behavior or sensory inputs.

Because interactions between neurons can occur anywhere from the dendrites to the axon terminals but one tends to think of the "activation function" as being applied at the spike initiating zone (the axon hillock), discriminating feedforward and feedback effects is often counterintuitive. For example, mitral cells, the secondary sensory cells in the mammalian olfactory system, form reciprocal dendrodendritic synapses with granule cells in the same anatomical layer (Shepherd 1990). When a mitral cell is excited by the olfactory receptors, excitation passes to the granule cell and, with a short delay and no intermediary spikes, inhibition returns, tending to end the mitral cell burst. These dendrodendritic synapses show plasticity and mediate rapid learning of odors in some social interactions; mitral cell responses to familiar odors are modified (e.g., Kendrick et al. 1992). Formally, one can say that recurrent pathways to the mitral cell change the dynamics of its output, but these recurrent paths do not involve the action potential mechanism.

Alternatively, one could say that the dendrodendritic interaction provides a parallel path with a delay element in a feedforward path to the mitral cell's action potential mechanism and this path just happens to include the mitral cell's own dendrites. Instead of decomposing the anatomical modules into additional units, it is easiest to speak simply of lateral interactions.

6.4 *Structured Recurrent Networks as a Substrate for Representing and Processing Data*

In the previous sections we described the rich dynamical behavior of layered networks with recurrent and lateral connections. Thus, it is not surprising that these networks may have stable states with locally well-defined activity patterns and that these states, as well as the associated activity patterns, may be changed by signals at the input layer.

An interesting and useful subclass of networks emerges as soon as we treat these stable states as discrete representations of the possible values of a variable. The actual value can be represented by the position of the activity peak or by the position of a sharp border between active and inactive regions in a topographically organized network. Obviously, the precision of the representation in this coarse code is limited only by the number of states allotted to the range of values to be represented. Long-term stability is another advantage of such "place" codes that cannot easily be achieved by analog neural coding, that is, those based solely on potential or discharge frequency. Neural maps can be circular, allowing periodical variables to be mapped uniquely and without the discontinuity that necessarily occurs in analog representations, for example, in going from 360° to 0°.

In this type of topographic network, small changes encoded in the variable correspond to small changes in the activity pattern or, equivalently, transitions between neighboring states of the network. In simulations (Hartmann 1992), a spike or a burst of spikes at the input layer is able to shift the activity pattern in a topographically organized network from one stable state to a neighboring stable state. Repeated spikes or bursts of spikes shift the pattern in incremental steps, corresponding to an incremental change in the value of the encoded variable. Due to this mechanism the integral of the input rate over time can be directly and precisely read from the activity pattern. Accordingly, networks of this subclass are called "integrating networks".

In many biological systems, sensory information related to the rate of change of a variable, for example, velocity of a limb, is encoded proportionally in the rate of action potentials. This rate, applied at the input layer of an integrating network, will shift the activity pattern at a rate proportional to velocity. Hence, the position of the activity pattern will correspond to the integral value of the velocity signal, which in the example is the momentary position of the limb. Integration of negative input values is possible if positive and negative velocities are encoded by separate channels that shift the activity peak in opposite directions.

Integrating networks with travelling activity patterns can be realized using a variety of different architectures compatible with those found in other recurrent networks. Accordingly, current anatomical and physiological data is not sufficient to determine whether integrating networks are implemented in biological systems. On the other hand, simulation results in the fields of neural motor control, spatial orientation and navigation are very convincing. Hartmann (1992) presented a model describing spatial representations of objects with respect to an observer's moving frame of reference. The necessary three-dimensional transformations are performed by realistic proprioceptive signals – which represent components of velocities and angular velocities – acting on two-dimensional integrating networks. Hartmann and Wehner (1995) incorporated integrating networks in a model of the navigation system of the desert ant *Cataglyphis fortis*. The model describes quantitatively all the behavioral results concerning the highly intelligent performance.

6.5 *Networks with Memory and Learning*

The networks considered above can generate complex transformations and sequences of behavior. The transformations can include complex, preprogrammed adaptations to changing circumstances; as already mentioned, considerable adaptation can be achieved by a feedforward system designed with sufficient foresight. The networks can have state, in the sense that the presence of a CPG or recurrent connections causes the network to react differently to a given input at different times (but this is not a plastic form of memory). Thus, such networks still lack one important feature of intelligence: the ability to change the transformation on the basis of experience.

Section 4 outlined the complexity of learning in animals and its basis in physiological and morphological plasticity. The conclusion was that learning is an expression of the dynamics of the interaction between an agent and its environment and these dynamics reflect plasticity arising through many, more or less closely linked processes acting with different time constants. A similar variety of learning algorithms have been developed and applied to artificial neural networks (ANNs). Many are modeled on biological processes and yet no approach captures the full range and complexity of natural systems.

Thinking of learning as a change in behavior resulting from experience seems to imply a recurrent process, one modifying the state of the network on the basis of some post hoc evaluation of signals and changes in state. As illustrated in Section 4, this need not be true for simple forms of learning like habituation and sensitization. In animals these can be performed by intrinsic processes in a single synapse and by feedforward interactions of two elements, respectively. The *Aplysia* results show further that even simple associative learning can be performed in a feedforward way, utilizing traces of preceding activity. (Recurrent signals may be involved at some level, even if only in the form of messengers to match growth in the pre- and postsynaptic elements). For artificial systems, in contrast, even such simple kinds of learning are usually performed by a separate system or module that monitors and evaluates the activity and adjusts the network accordingly. In other words, there is a separation of learning and execution processes not evident in animals.

In another respect, work on artificial neural networks has followed animal models. Both kinds of plasticity evident in animals have been implemented: short-term changes in the connection strengths and long-term changes in the number of units and the connectivity. To the extent that these methods improve performance, they imbue the network or animat with increasing intelligence, at least intelligence of design. Rather than discuss individual learning algorithms in detail (see Braun, this volume), we will consider only their relationship to intelligence, beginning with the short-term changes.

Several general approaches have been applied to determine connection weights (Hertz et al. 1991). One is supervised learning, or learning using examples of correct input-output pairings (e.g., the back-propagation algorithm). This is simply a design tool and does not imbue the network with intelligence because the system has little autonomy. A second approach, that of unsupervised learning, does convey some autonomy or at least makes the control by the designer more indirect. In one kind of unsupervised learning (e.g., Kohonen networks) the designer provides only the rules for changing the network. These rules are chosen in such a way that they extract patterns present in inputs. In a second kind, reinforcement learning, the designer also provides a qualitative or quantitative measure of performance. In this case, the rules for changing the net may simply implement a random search, or they may produce changes dependent on the preceding pattern of activity in order to use added information that one hopes will increase the speed of learning. In either case, the evaluation function constrains the agent's autonomy even if it may not constrain the final structure of the net. If the network is prestructured appropriately (see below), successful unsupervised learning can develop implicit or explicit, context-specific models of the world.

Long-term changes in networks may also involve introducing new units and new connections. Here, the natural model is the mutation and selection mechanism of natural selection; several variations have been proposed. Alternatively, neurons can be added or subtracted systematically depending on some performance measure for the network as a whole. In artificial systems, neurons are simply added with random connections. In animals, neural growth within individuals is affected by the neuron's activity, so it is conceivable that the generation or loss of connections is not entirely random but directed, at least with respect to which neurons are selected for growth. Across generations, however, the introduction of new neurons seems to be a matter of chance mutation and not modified on the basis of experience. In any event, the adjustments in the substrate of intelligence would appear to represent an intelligence of process unrelated to any intelligence of content.

In summary, theoretical considerations and biological examples suggest that significant features of agents and natural environments change with different time constants, so both short-term and long-term changes in networks are necessary for mastering complex tasks in changing environments and thus improving the intelligence of process in the agent.

6.6 *Modular Networks*

Artificial neural networks do not have the heterogeneity in types of units or, in many cases, the physical topology of real nervous systems, so modularity can only be defined in terms of clustering with respect to connectivity. Modularity can be introduced into any of the types of network architectures described above, either by design or in the course of training the connections in a totally interconnected net. In the former case, it is a way to simplify the analysis and maintenance of complex systems (see Ritter, this volume) by constructing modules along functional divisions. This may also be an efficient architecture on other grounds, for example with respect to physical connectivity. In the latter case, emerging modularity may fortuitously aid understanding of a natural or artificial network, but certainly in biological systems, developmental and historical constraints may mean that modules defined by connectivity do not strictly correlate with functional subdivisions.

The most general type of network would be a fully connected recurrent network. Modularity introduces order into such a network, so one can ask whether new characteristics emerge as a result. Modularity by design in artificial systems can be hierarchical or parallel in accord with functional relationships.

Hierarchical modularity by design is appropriate when a task can be broken into a series of steps or separate functions. In this case, individual nets can be trained to perform parts of a complex transformation, so that the output of one net provides the input for another in the final system. An example is the separation of parts in reinforcement nets using adaptive critics (e.g., Barto et al. 1983). In this case, the net is structured and the learning rules are organized so that one part learns a transformation and a second module learns an evaluation of the transformation currently used by the first module. In this way, the network gains autonomy of organization, but any intelligence of content is still in the designer.

In other cases, modules may be more or less equivalent, but because they have to compete for the same motor apparatus, there must be a mechanism for deciding when a given module affects motor output. For example, the problem of coordination arises when two or more CPGs are implemented in one system. These may be the control systems for discrete behaviors (e.g., feeding, mating, fleeing, etc.) or those for subsystems contributing to a particular behavior (e.g., those for the different limbs that participate in walking).

Obviously, there is a problem of definition here similar to that involved in identifying modules: one could consider each wing-beat generator for a single wing as a separate CPG and examine the interaction among the four CPGs or consider the network for all four wings as a single CPG. Often boundaries are drawn as a matter of convenience for a particular experimental focus. In the walking system of the stick insect, for example, we expect that connections among the neurons controlling a single leg will be more numerous and complex than those between neurons controlling different legs, simply due to the separation and required conduction times between ipsilateral legs. Behavioral evidence confirms that each leg has its own step pattern generator. Coordination of the legs emerges from numerous mechanisms involving signals exchanged between these step pattern generators (lateral connections) rather than mediation at the level of the motoneurons (integration of multiple inputs). As in earlier examples, this architecture lends the insect intelligence of process without involving any explicit logical processing. This intelligence is expressed in the presence of variable, adaptive step patterns, like those of relative coordination (v. Holst 1969), rather than a single, rigid coordination.

7. COLLECTIVE INTELLIGENCE

Societies of individuals are complex systems that may show a particularly clear form of modularity. As a unit, a society or group can exhibit intelligent behavior. For such cases, the relationship between intelligence of process and intelligence of content is often clearer than for intelligence in individuals, the primary subject of previous sections. Collective intelligence can be defined as intelligent behavior arising from information processing and problem solving based not merely on the capacities of individual animals, but rather on the interactions among animals belonging to a group. In the extreme, collective intelligence is a decentralized process emerging from interactions between distinct, autonomous units that individually sense only local stimuli and exhibit behaviors that are simple and probabilistic. Each unit responds with specific behaviors (attraction – repulsion, activation – inhibition) to local signals emanating from the physical environment and to signals emitted by other agents. Several examples of collective intelligence will be explained here in more detail because they may provide useful paradigms for viewing neural networks. In this analogy, individual animals, which are readily observable, play the part of neurons, which are less easily monitored.

The clearest natural examples of collective intelligence are from the behavior of social insects (ants, bees, wasps and termites). Because collective intelligence depends on cooperative and/or collective action, which often means that some individuals give up reproductive potential, the evolution of such societies was early recognized as a problem for evolutionary theory (Darwin 1859). However, the analysis of conditions under which selection can work at the level of the colony rather than the individual has resolved this apparent contradiction. In particular, the prevalence of highly social forms within the *Hymenoptera* has been linked to the special kinship relationships within this group (Hamilton 1964). Further research has uncovered several adaptive mechanisms contributing to the collective intelligence shown in many features of complex insect societies.

An early line of research considered the internal organization of insect societies; it focussed on the determinism of individual behavior. These studies assumed that the increasing complexity of the tasks that must be performed in a colony is reflected in an increasing behavioral specialization of the workers (Wilson 1971, 1975; Oster & Wilson 1978; Wilson & Hölldobler 1988). This assumption is based on the observation that the behavioral repertoire of individuals from a social species is not more extensive than that of insects from related non-social species. In fact, individuals from highly social species often appear to show less flexibility than those from less highly social species. Consequently, the evolutionary success of social insects is not caused by innovation in the behavior of the individuals, but relies on the division of labor between reproductives and workers and on the behavioral specialization through differentiation of morphological and/or behavioral castes. As each group is responsible for only some of the required tasks, different tasks can be performed in parallel. The large number of workers performing a task makes the system tolerant with respect to individual failure or error. In other words, despite a relatively high error rate in individual behavior, parallelism increases the success rate of the behavior at the colony level. Thus, this model of insect societies is based on individual determinism and specialization through differentiation. It suggests that increasingly specialized functions on the colony level are based on increasingly specific adaptations of its members. In this view, the occurrence of collective behavior and structure is the consequence of predetermined behavioral programs in the individuals.

A second line of research, motivated by the work of the Brussels school studying self-organization in physico-chemical systems (Glansdorff & Prigogine 1971; Nicolis & Prigogine 1988), sheds a slightly different light on the spatial organization of social insects, as expressed in building, foraging and recruitment (Deneubourg et al. 1984, 1986, 1987; Pasteels et al. 1987a,b). The underlying concept is that the social organization of insect colonies is not predetermined by genetic mechanisms; instead it emerges from the continuous interactions of a large number of individuals that exhibit only simple, stochastic behavior and possess limited knowledge of their environment. The principal mechanism controlling the spatial organization of the colony is a positive feedback loop that can amplify small variations in the behavior of individuals to cause changes at the level of the colony. The positive feedback arises because the workers communicate with each other. The dance language used by bees to indicate direction and distance of food sources is an elaborate form of communication, but most communication in social insects is based on pheromones, simple chemical signals that carry a simpler message. The major point here is that a chemical message changes the behavior of nestmates in such a way that others are more likely to repeat the behavior and the pheromone emission performed by the sender. Thus, the chemical communication has an autocatalytic nature that can transfer to the collective level small variations in individual behavior that initially are present only locally. Therefore, the collective behavior is more than simply the sum of individual actions; it shows new characteristics that emerge from the self-amplifying interactions among individuals and between individuals and the environment. This collective behavior is self-organized and qualitatively different from the individual behavior. As in the first model, the individual behavior may be variable and inefficient, but these deficiencies are compensated by the reliability of the collective behavior. Thus, the model proposed in these studies exploits the self-organized behaviors of a group of interacting agents to enable collective behaviors exceeding the complexity of the individuals.

7.1 *Collective Decision Making as an Intelligent Process*

In observing an ant colony, one is often astonished by the apparent inefficiency of individual workers. One worker may carry a food item towards the nest and drop it somewhere nearby, only to have a nestmate pick it up and carry it in the opposite direction. Nevertheless, food is gathered, broods are reared, a highly complex nest is constructed and the colony does survive and reproduce. Moreover, the overall activity of the colony seems to follow a well-defined plan and to make "choices" reflecting global information about the environment. How can complex patterns be generated from the individual behavior and what are the rules coordinating numerous "agents" in order to produce a useful structure out of an initially chaotic behavior? We would like to consider these points using three examples.

The first concerns food recruitment and the selection of food sources. In ant societies, food recruitment is a collective process mediated by chemical communication. Initially, the active foragers randomly explore the foraging area in a search for food. When a worker has found a food source, it returns to the nest, marking the substrate with a chemical trail as it goes. The ant modulates the strength of its marking according to the quality of the food source. The trail pheromone incites nestmates to leave the nest and follow the trail towards the food source. However, trail-following is not precise. Many ants lose the trail, particularly at the beginning of recruitment when the trail is weak. However, recruits that do find the food add their own pheromone in returning along the trail to the nest and in this way reinforce the chemical marker. This, however, has an important consequence because the reliability of trail-following increases with pheromone concentration: as recruitment to a good source proceeds, the concentration of trail pheromone increases, which increases the number of individuals successfully reaching the food source. This positive feedback leads to a structured, collective behavior in which the colony concentrates foraging activity on only one part of the foraging area. The initially random search is progressively restricted until the majority of foragers move along a single path to the food source.

When two or more food sources are present in the foraging area, the same mechanism leads to collective decision-making. If the sources differ in quality, the average amount of pheromone deposited by returning foragers differs accordingly, changing the dynamics of recruitment. The pheromone concentration of the trail leading to the richest food source increases faster than that of the others and, consequently, attracts more foragers until it becomes irreversibly dominant. Thus, the appearance is as if the colony had "selected" the richest food source, even though no single ant makes an explicit comparison of the food sources and there is no central decision instance.

Even when the food sources are equally good and all are initially visited by approximately the same number of ants, small chance variations can result in one trail becoming slightly stronger. The positive feedback then reinforces this difference, leading to the preferential exploitation of only one of the food sources

Another related example mediated through chemical recruitment is trail optimization. In an experimental situation the nest is connected to a food source by a bridge with two branches. The branches can have equal or unequal lengths. By a process analogous to that described above, the colony establishes a preference for one branch, usually the shorter of the two. When recruitment begins, the ants explore using both branches. If one branch is longer than the other, the foragers concentrate more and more on the shorter path until nearly all of the individuals use it. This is due to the fact that if the choice of branches is initially equal, the shorter travel time and the higher percentage of ants completely traversing the shorter path result in a stronger pheromone concentration. Besides this collective mechanism based on trail-laying and trail-following behavior, individual orientation preferences contribute to the selection process. Foragers traveling along a branch make U-turns when the path deviates strongly from the straight line between nest and source or when the pheromone concentration suddenly drops because the forager has chosen the less-used branch. Thus, redundancy is present. However, the essential part of the selection process is the positive feedback by which small initial differences are amplified to guide the large majority of the foragers along one of the alternatives. If the branches are equal in length, random fluctuations can be amplified, causing a preferential use of only one branch. If the branches are nearly equal in length, the same stochastic effect can cause the trail to become established on the longer branch.

The third example is nest building in termites. In 1959 the French biologist P.P. Grassé proposed the theory of "stigmergy" to account for the construction of complex termite mounds. When termites are placed on a homogenous surface covered with moist soil, they immediately begin nest building. In this behavior they form little soil balls with their mandibles and impregnate them with saliva containing a pheromone. The pheromone is attractive to other termites, increasing the probability that they will approach and add their own soil balls to existing piles. Thus, the termites initially carry balls and drop them randomly, producing irregularities in the form of small aggregations of various sizes. However, because the probability of depositing a ball is related to the quantity of pheromone, which in turn is proportional to the size of the pile, the bigger aggregations grow faster than the smaller ones and form little columns. When the columns reach a certain size, the sphere of attraction extends to termites on neighboring columns, causing them to build preferentially toward nearby columns. In this way, columns become linked into arches, which then form the foundation for a new construction. Like the previous examples, this phenomenon relies on the stochastic behavior of the insects on the one hand and on the amplification of local inhomogeneities on the other.

7.2   General Characteristics of Collective Intelligence

In all three examples, the behavior of the colony as a whole exhibits many features characteristic of intelligence, including the following three:

Flexibility: A change in the environment is perceived first on the individual level and then, through amplification by positive feedback, on the collective level, so the system adapts spontaneously to new circumstances. This adaptability eliminates the need for extreme individual specialization and preprogramming for all possible situations. The problem posed by unforeseen situations is eliminated because comprehensive, a priori, global information is not needed.

Complexity: The colony behavior adapts to quite complex environmental situations and, of course, also solves the difficult problem of coordinating the immense number of individuals making up some ant societies.

Reliability: Due to the large number of identical agents, the system can still function in a normal way if some units fail – even without any centrally organized reassignment of duties.

The behavior of a single worker is not sufficiently complex to deal efficiently with these problems. Moreover, the individual would need complete knowledge about the environment and the task, which it obviously does not have. (The perceptual horizon of an ant is approximately 1cm!) To adapt to this situation, social insects have shifted some complexity from the individual to the interactions between individuals, allowing the colony to exploit the stochastic behavior of its members. Self-organizing structures bridge the gap between the size of the exploited area on the one hand, and the size and perceptual range of the individual on the other hand; they restore to the colony a behavioral flexibility given up at the level of the individual.

The collective intelligence in the decisions is the result of several autocatalytic processes that evolve with different dynamics depending upon environmental conditions (the "richness" of a food source, the geometry of the branches, the perception of the local environment modified by building activity, etc.) and the individual's reactions to them. Note that each individual can only perceive and react to local stimuli and that it has no global information about the environment (the location of food sources, which of the two branches is shortest, the location and shape of the new nest). These features can be characterized in the following behavioral rules:

1. Each agent perceives only local stimuli and, consequently, has only limited knowledge of its environment.
2. Local stimuli trigger in each agent a behavior that is modulated according to the characteristics of the local situation: for example, ants vary the intensity of the emitted communication signal according to food quality. Physical and geometrical constraints directly affect the propagation of signals and consequently influence the individual responses and the emerging collective behavior. These circumstances allow the generation of adapted responses even without a symbolic representation of the environment or a highly specific communication.
3. Each agent exhibits probabilistic behavior. The probability of performing a certain behavior is correlated with the intensity of the stimulus. This feature, which explicitly incorporates errors on the part of the individual, is a necessary condition for the colony as a whole to shift from one pattern to another.
4. The interactions between agents generate positive feedback mechanisms. Each agent may change the behavior of another one in the sense that it increases or decreases the probability that the second agent will perform the same behavior as the sender.

The reliance on self-organization shows that adaptation to variable environments in social insects is not necessarily achieved through morphological adaptation of specific castes. When behavioral flexibility is required, adding complexity based on interactions of simple agents constitutes a feasible alternative to increasing individual complexity. The optimization via the interactions between agents generates a collective efficiency that is beyond the capacity of each individual. Note that more "intelligent" agents do not necessarily increase the group's efficiency. More important is an optimal balance between individual capacities (the behavioral rules of each agent) and collective capacities (self-organized behaviors on the group level).

More generally, the examples show how a distributed system of simple agents can find solutions to problems without explicit coding. Instead, solutions emerge from the interactions between agents on the one hand, and between agents and the local environment on the other. Problem solving in the present context becomes a morphogenetic process, a kind of pattern formation where the reorganization of the agents and the group reaches a final configuration – the solution – consistent with the physical constraints of the environment. Extensions of this concept to groups of robots are discussed in the following section and in the chapters by Lanz and McFarland and by Ritter. In both ants and robots, the intelligence of process clearly arises without intelligence of content.

The examples discussed here and in the previous section show that cooperation of many "simple" elements – be they individual neurons, separate modules of many neurons within a brain, or individual animals forming a superorganism – can lead to an immense variety of behaviors that exhibit many properties of intelligence in the absence of rational thought, or in other words, exhibit prerational intelligence. In many cases, a naive observer ignorant of the underlying mechanisms might even have the impression that the system has insight (intelligence of content). How such a system could be endowed with high-level cognition, the ability to manipulate symbols and reason logically, will be discussed in Section 9.

8. INTELLIGENCE OF ARTIFICIAL SYSTEMS

The properties of artificial systems built in the field of Artificial Life, often called animats, are discussed in detail (see in particular the chapter by Ritter). In this section, we want to briefly consider these developments from the perspective of biology. Two related points are of interest. First, what is the relevance of this research for biologists seeking to understand how animals and biological systems function (Dean 1998)? Most work to date relates to this question. Second, what are the similarities and differences between animals and artificial systems and, in particular, what is the relevance of artificial systems to understanding natural intelligence? The latter point includes the question of how autonomous systems such as animals could have developed intelligence and cognition without the aid of an external, intelligent designer.

With respect to the first point, we strongly believe that work on artificial systems complements traditional biological research in important ways. Systems or organismal biologists want to understand how biological mechanisms work as a whole. This means understanding the structure and function of sensory systems, nervous systems and motor systems as well as their interaction with the environment and mechanisms of adaptation and learning. In this respect simulation is a necessary tool because many, even relatively simple properties of animals and their nervous systems cannot be fully understood if only analytical, reductionist methods are used. Synthetic methods like simulation are necessary in order to obtain an idea of how the independently investigated mechanisms combine and interact. In particular, simulation can provide insight into how new properties can emerge through interactions.

A general, methodological question, but one with important implications, is whether complete, autonomous systems should be constructed as pure computer simulations incorporating a simulation of the environment or as physical robots "situated" in a real environment. Previous sections have emphasized that the details and physical properties of the environment play a crucial role in the interaction of agent and environment, so an adequate simulation of this environment is necessary. However, this can be very difficult or impossible; it is easy to overlook features that may impede or facilitate useful behaviors. For these reasons, simulation of whole animals based on situated robots is the ideal goal, although often computer simulation is a sensible first step. Biologists do not usually have the know-how or the facilities for constructing real robots, so this is an important area for interdisciplinary collaborations.

8.1 *Simulations of Specific Biological Systems*

One group of simulations is of immediate relevance to biologists: those that have arisen out of biological research and seek to understand a specific biological system. Such simulations synthesize individual experimental results and can be used to develop hypotheses and predictions that can be tested in further experiments. This is particularly true of attempts to realistically model neurophysiological and anatomical data on biological neural networks, an important area that will not be considered further here.

Several examples illustrate this approach as applied to the behavior of whole systems. The first two concern sensory capabilities and their use for controlling movement. The simulations of Hartmann and Wehner concerning sun-guided navigation by ants have already been mentioned in Section 6. Franceschini et al. (1992) built an analog hardware simulation of the fly retina, including both photoreceptors and neural movement detectors. By mounting this sensory system on a mobile robot, they could show that this passive system of multiple elementary units was sufficient to support obstacle avoidance behavior. Simple summation of signals from obstacle detectors and from a detector for direction to the destination was sufficient; cross-channel inhibition or suppression of signals was not necessary. The non-uniform arrangement of the photoreceptors observed in the fly and implemented in the robot was found to simplify the computation of the optical flow field, a fact that may generalize to many other animals including humans. Finally, the system detects obstacles only when the robot or the obstacle is moving, but if neither is moving, avoidance behavior is not necessary. This is a good example of the importance of realistic interactions with the environment.

In similar vein, Webb (1994) constructed a robot that mimics the phono-taxis of female crickets attracted to a male's calling song. Her robot demon-strates that phonotaxis is possible without a hierarchically superior module for explicit recognition of the species-specific song. Thus, the temporal pat-tern of the calling song does not need to be encoded and discriminated, as was generally assumed. The model suggests that physiologists should fo-cus on characterizing neurons with specific directionality properties rather than searching for neurons that encode temporal features of the sound. The simulations in a real environment also revealed that simply summing the phonotactic signal with obstacle avoidance signals sometimes left the robot trapped behind obstacles, so non-linear interactions are necessary.

A simulation of the walking behavior of a six-legged insect provides an example focussing on motor systems, although the sensory elements in the simulation are also important. These computer simulations and a robot im-plementation conclusively showed that combining the coordinating mecha-nisms identified in many different experiments does lead to consistent, nat-ural behavior of the whole system (Dean 1991; Müller-Wilm et al. 1992; Cruse et al. 1995). Further, several questions resulting from the hardware simulation led to new experiments. These questions arose specifically out of the interaction between machine and environment; they refer to the role of load sensors, the problem of coping with difficult initial leg configurations and the role of positive and negative feedback – both of which are present in the insect.

Several beautiful simulations concern the collective intelligence shown by groups of individuals (Pasteels 1987a,b). The recruitment behavior of many foraging ants and its dependence on positive feedback involving a chemical trail marker were discussed in Section 7. Experiments showed that when ants discover two sources of equal quality, they initially exploit them equally, but at some point a symmetry break occurs and most animals visit only one of the two sources. Simulation results supported the hypothesized mechanism described in Section 7 and showed further that the outcome also depends on the size of the colony. For very small colonies, no trail devel-ops. Colonies of medium size develop trails, but the symmetry break does not occur and both sources are exploited equally. The symmetry break only occurs for large colonies. These predictions can now be tested experimen-tally. Theraulaz et al. (1991) used simulations of wasp colonies to show how a few simple rules can produce different functional groups – that is, queen, foragers and nest keepers – by a process of self-organization.

Successful simulations of specific systems, like those described above, help us to understand emergent properties that arise from the interactions of different subsystems. They also test whether the experimentally confirmed mechanisms are sufficient to explain the observed behavior of the whole system. When known mechanisms turn out to be insufficient, simulations can lead to new questions for behavioral experiments. Simulations, particularly those based on artificial neural networks, can also provide hypotheses and guidelines for designing and interpreting neurophysiological experiments.

8.2   *Simulations Not Tied to Specific Biological Systems*

A second group of simulations is more oriented towards engineering and general design principles not tied to specific biological examples. However, these studies can also provide new insights for biologists. Two directions are particularly interesting. One is the investigation of different architectures and the question of how different functional modules can be arranged in order to produce sensible behavior in the complete system. The other concerns the question of autonomy. This addresses the nature of the mechanisms by which living systems without the aid of an external designer develop during evolution and ontogeny the ability to survive and to cope with fairly unpredictable environments.

Rodney Brooks (1986, 1991) addressed the first question and proposed a very simple principle, the subsumption architecture. This principle involves, first, an analytical decomposition of behavior into smaller behavioral units that can be performed by simple finite-state machines and, second, the combination of these units in a logical hierarchy of control. The building blocks in the subsumption architecture and related approaches – modules performing a well-defined behavior – are behaviors and the approach and the architectures are referred to as behavior-based. More generally, a behavior can be defined as a regularity in the dynamics of the interaction between agent and environment (Smithers 1992). Because this approach builds from small behavioral units, it is radically different from classical AI in which all system knowledge is united in a world model and all sensory inputs must be processed in the light of this model before actions occur (e.g., Krotkov 1991). For an engineer, the subsumption structure has the great advantage that it can be more easily analyzed. For the biologist, it resembles biological systems in that many basic modules are simple, but it differs in that the strict separation and logical arrangement of modules seems at odds with the opportunistic and heterarchical appearance of biological control networks. Astonishingly complex behaviors can be produced with this architecture, which suggests that it captures essential features of many behavioral tasks. Nevertheless, the explicit hierarchical organization of subsumption architectures limits their versatility.

In answer to this problem, Maes (1991) extended the behavior-based architecture by relaxing the hierarchical constraint. In her model the hierarchy is only implicit and it is soft in the sense that the functional hierarchy of behavioral modules depends not only on the predefined structure of the system but also on internal and external factors, such as autonomous intrinsic changes and sensory inputs. Some of these ideas are very much related to early proposals by the ethologist Lorenz (1950, 1965) and the biocyberneticist Hassenstein (cited in Eibl-Eibesfeldt 1980). Again the interesting result for the biologist is that, although the internal structure is quite simple in principle, such animats can show unexpectedly complex behavior, including displacement activity, attention shifts, and opportunistic behavior reflecting the balance between sensory-driven responses and motivation or goal-driven behavior.

8.3 *Relation to Intelligence*

In all the approaches described above, the artificial systems were developed by an intelligent designer. A critical question is to understand how such behaviors might develop autonomously, that is, without such a designer. Experimental approaches to this question have focussed on the emergence of new behaviors through combinations of simpler behaviors and interactions of simple modules. "New" in this context means at the very least not explicitly designed into the artifact.

Several examples involve emergent behavior in fixed-rule, behavior-based systems. Smithers (1992) showed that a machine designed with the two behaviors "turn left" and "avoid obstacles" would exhibit behavior describable as "wall-following". Similarly, the robot of Beckers and Holland, discussed in the chapter on philosophical perspectives, has three behaviors: *(i)* turning to avoid obstacles, *(ii)* pushing small objects forward, and *(iii)* stopping, reversing and turning – which releases any objects being pushed – if the frictional force developed by the pushed objects exceeds a given threshold, which means that the number of pushed objects is too large. As a result, the robot gathers objects in its workspace into a single pile. This "collecting" behavior and even recognition of a pile are not explicitly implemented in the system. Matarić (1992) gives examples of group behaviors like flocking or homing emerging from a few simple behaviors implemented in the individuals. All these systems are essentially feedforward, reactive robots with no plasticity. Although they do exhibit emergent behaviors, these behaviors are still fairly primitive.

Verschure et al. (1992) demonstrated a simple system that could develop new sensorimotor connections. The robot was equipped with tactile sensors and a reflex that permitted it to move around obstacles detected by contact sensors. It was also equipped with visual sensors that respond to obstacles at a distance, but the connection strengths between these sensors and the motor output was initially unspecified. (The presence of some connections, of course, provides the necessary structural basis for developing a functional connection.) Beginning with such a structure and the capacity for unsupervised, Hebbian learning, the robot learned to use its range sensors to avoid obstacles from a distance, that is, without bumping into them. The presence of plasticity and feedback in the form of an evaluation function make this system a step in the direction of more intelligent systems, but the behavior is still rather simple and one might say that the game was rigged from the start – via the evaluation function – to obtain the desired result.

Simulations of collective intelligence represent an extension of the behavior-based approach, so they will be mentioned here although they are closely related to the studies of ant behavior discussed in Section 7. The essential feature is that, like the other examples, the collective system demonstrates intelligence of process in the explicit absence of intelligence of content. The plasticity in the collective behavior resides not in designed plasticity in the individuals but in the dynamics of the interactions among the individuals. This approach also disavows the classical AI ideal of systems of agents with highly developed representations and knowledge of the environment. It asserts that a group of reactive agents with only simple behaviors and no representation or knowledge systems can exhibit highly intelligent behavior showing the following characteristics:

1. faster performance because a large number of identical agents contribute;
2. highly robust and reliable behavior;
3. qualitatively superior and more efficient performance than that of the individual agents.

Viewed as a problem for centralized, hierarchical control, the use of multiple agents to enhance performance speed incurs additional effort in terms of supervision and coordination. Centralized control may be tempting from the standpoint of optimization and it has been pursued in AI and engineering, but its use is restricted to certain situations. Control by a human operator is unrealistic; the difficulty of controlling the activity of a multi-agent system in real-time typically exceeds the capabilities of a human operator. Several general problems arise in using a central processing unit that continuously monitors all agents and distributes instructions to them. First, there is a total dependence on the reliability of the central computer and its peripheral systems. Second, the coordinates of each agent, needed for the central control, may be difficult to obtain due to sensor inaccuracies, signal propagation delays and limitations of channel bandwidths. Third, when confronted with unforeseen situations (the failure of one or several machines, the presence of obstacles, etc.) a centralized control system requires highly complex algorithms to handle the explicit coding of all possible situations.

In summary, several real robots and collections of real robots show autonomy and various emergent behaviors on a small scale. For the biologist as well as the engineer an important and as yet unsolved question is how these principles will scale up to larger systems. The examples described to date contain relatively simple internal structures and the behaviors are simple. Merely increasing the number of units in networks is surely inadequate, due to the increase in search time needed to find workable systems. Do additional principles have to be invented? Modularity promises some relief, but if the system consists of many modules, the question arises as to how connections between these modules might be determined automatically and how they could be selected for learning. Sometimes it is argued that "tabula rasa" approaches like genetic algorithms (GA) or reinforcement learning, which clearly are biologically inspired, may solve the problem. However, there are practical and theoretical objections. The practical problem is that the more complex and structured systems become, that is, the lower its entropy in a sense, the more time is required to search for and find a workable system in the absence of any preexisting knowledge. Animals, of course, have an immense amount of structure, representing implicit knowledge and algorithms, stored in their genes, which complicates comparisons with artificial systems. The theoretical problem is that both methods require an important input from the designer in the form of the evaluation function. It can be argued that this function implicitly specifies the desired output.

To conclude this section, one can say that the development of these architectures has provided some insights for biologists and certainly provides new tools for thinking about biological systems. However, we suspect that these architectures are more novel for engineers and AI-trained computer scientists. Many principles are already evident in biological nervous systems. Similarly, the behavior-based approach is quite consistent with much low-level processing in animals. However, it is not yet clear that it can be extended to capture the essence of the recurrent networks and bidirectional processing that is so evident in biological nervous systems. Exploring the capabilities of simple reactive robots using this kind of control is currently a very active area of research.

9. CONCLUSIONS AND OUTLOOK

This chapter began by considering different attributes of intelligence. That animals and humans show intelligent behavior according to these criteria was taken to be self-evident, so we did not feel the need to present examples for each criterion. The fact that behavior is adapted to promote reproductive fitness under many different environmental and social situations is sufficient evidence. Then, we considered features of animals that we believe to be important elements contributing to intelligent behavior. This review highlighted several features that all contribute to the high complexity of biological neural networks. The first is that the basic unit, the neuron, is highly complex in its physiology, morphology and biochemistry. Units in current artificial neural networks are much simpler. Second, neurons are heterogeneous and adapted for different functions. Units in artificial neural networks are much more uniform, although the advantage in adapting units to a task is evident, for example, when the choice of linear or logistic activation functions in the output layer can be matched to the nature of the transformation to be simulated. Third, biological neural networks are highly, but not fully interconnected and the networks are structured with a mixture of local and long-distance connections. Every neuron is not connected to every other one, simply for reasons of space. Fourth, plasticity occurs on several different levels with several different time scales. Plasticity does not depend on a distinct "plasticity" or "learning" module; instead it emerges from the integral workings of the cell. Different influences converge onto several different second messenger systems within the cell and then fan out to affect different, but partly overlapping, aspects of neural function. Thus, the neurons and the systems of which they are a part are dynamic systems retaining various traces of their inputs and outputs that in turn modify future activity. The capability for plasticity is heterogeneous: some synapses are highly plastic, others are constant. Fifth, nervous systems of higher animals show considerable modularity on several different levels. Sixth, many biological circuits and procedures appear to be ad hoc constructions and working approximations. Evolution rewards performance first and design only indirectly via performance. This penchant for using anything that works reasonably well is different from the exact calculations epitomizing technical solutions, although we may not appreciate how many everyday engineering solutions are simply working approximations.

Because present-day biological networks, even in simple organisms, show a mixture of features, that is, they do not fit into a clear evolutionary progression, we next considered different architectures for artificial neural networks together with the intelligence they might exhibit. This review establishes that even simple systems can show unexpectedly complex behavior, which is one aspect of intelligence. These simple systems can show some elements of cognition, but not others, as will be discussed below. The survey further showed that with increasing complexity of structure increases in different attributes of intelligence do not necessarily advance at the same rate. Of many attributes, learning – the ability to change behavior based on experience – is a critical prerequisite for intelligence.

The conclusion is that systems based on prerational mechanisms exhibit a gradation of intelligence. The examples concern behavior involving intelligence of design but not of content. The behavior is intelligent in that it has utility according to the purposes of the designer. The behavior may be emergent in the sense that it is not explicitly specified by the designer but arises through the interaction of implemented processes. Regardless of how complex the behavior is, it is nevertheless predictable, at least in theory, because it follows the rules and procedures implemented by the designer.

Does this gradation of intelligence continue to higher forms of intelligence evident in humans? In some contexts, intelligent behavior, like intelligence of design, leads to very predictable behavior. In others, an intelligent insight appears suddenly and thus often seems unpredictable and nondeterministic. Does intelligence of content require a qualitatively different set of procedures? A belief in the continuity of biological evolution and the similarity in the elemental and organizational principles of nervous systems in humans and other animals both suggest a negative answer. Instead, the phenomenon of higher forms of human intelligence must rest on procedures common to other animals, where we believe that only intelligence of process and intelligence of (evolutionary) design are present. Nevertheless, evolution does provide examples of radical breaks, such as new morphological features that open up new niches and lead to rapid evolution and radiation, so the evolutionary argument is not conclusive. Therefore, a useful alternative is to consider how intelligence of content might be realized in artificial systems.

At the outset, one should remember that there is a suggestive link between type of mechanism and type of behavior. By this we mean that intelligence of process seems to relate naturally to intelligent behavior where the behavior itself is a procedure (i.e., procedural knowledge or learning), whereas intelligence of content seems to relate naturally to intelligent behavior involving declarative knowledge or learning.

Recall that procedural knowledge, "knowing how", is connected to a given procedure or a given action (e.g., riding a bicycle or playing a piano). This type of knowledge can be used by the system only in this procedural context. Thus, it corresponds to a situation where a functional module has its local memory that cannot be used by others. In contrast, human declarative knowledge, "knowing that", by definition can be adequately and sufficiently expressed in words. A fact or experience or logical relationship can easily be reported verbally. Although one can try to explain in words how to ride a bicycle and one can even give a formula for steering angle as a function of speed and inclination, one is immediately aware that this description does not really convey the essence of the skill; it does not convey the ability to reproduce the same level of performance. Declarative knowledge, unlike procedural knowledge is not tied to a particular process or to a context closely related to that in which it was obtained.

As is often the case, the procedural-declarative distinction is not absolute; there is some transferal between the two, as illustrated in the bicycle example. On the one hand, declarative, symbolic information can be communicated and used for modifying procedural knowledge, as in teaching a sport. On the other, the processing of symbols and logic must rely on procedures, whether innate or self-organized in the course of experience.

The important point for our discussion is that declarative knowledge appears less tightly restricted to a particular task or domain; it can be manipulated and used in new contexts. This manipulation of declarative knowledge equals cognition, according to McFarland (1991); it is called "minimal cognition" in the chapter on Philosophical Perspectives (Lanz & McFarland, this volume). Thus, a system has (minimal) cognitive abilities if it can deal with stored information independent of the context within which this information was acquired. As before the distinction is relative and depends on just how one delimits a context. Nevertheless, declarative facts and relationships can be used in a wide variety of different conversations and contexts whereas a particular procedure, like a tennis forehand, may be applicable to other racquet games but not much else. Manipulation of knowledge is clearly related to higher forms of cognition involving awareness and judgement as well as simply knowing (see below).

With this distinction in mind one can say that the approach originally advanced in artificial intelligence attempted to solve declarative types of problems using systems with intelligence of content. These studies focused on problems involving symbolic representations and logical thought. The hypothesis was that a proper architecture for solving such problems was a system containing explicit factual knowledge and explicit logical operations. (When it came to processing input from sensors, as opposed to symbolic input, difficulties arose.) The totality of manipulable knowledge was called a world model. The important point was that this world model was envisioned as a substrate on which planning and hypothesis testing could be performed. Knowledge about relationships could be used to test the consistency of noisy, incomplete sensory inputs against templates for known objects defined in the world model. Knowledge about possible outcomes of actions could be used for planning what action to perform. The important point here is that an explicit world model decoupled from specific procedures could allow different actions and their probable consequences to be tested mentally before actually deciding what action to perform. For a sequential processor like a Turing machine or a von Neumann computer such a serial evaluation of options means that the process requires a finite time before the system is able to act. Animals, in contrast, can respond at any time, even if the full specification of an action is not complete.

As an aside we should mention that intelligence of design could be said to include an implicit world model. Every network (and every module) has an implicit world model in the form of the weights connecting stimuli from the environment to appropriate actions of the system. As long as these actions are "appropriate", the set of weights represents an implicit world model that is valid but may be simplistic. A simple central oscillator or a biological clock may, for example, be interpreted as a world model that predicts changes in the environment.

Because we assume that biological systems have declarative intelligence and manipulate symbols using procedures like those used in non-declarative processing, we should consider what might be involved. Because the use of symbols and logical thought can be dissociated in development (Piaget & Inhelder 1977), we can separate these two capabilities and ask qualitatively how they might be realized in artificial systems.

Declarative knowledge, because it relates to language, necessarily involves symbols and relationships. What, however, does it mean for a neuronal system to store symbols? A symbol represents its referent in some way. However, a symbol is usually thought to provide a special kind of representation, one that does not store all the details but only a shorthand or generalized version of the item. In a sense, this transformation from specific instance to symbol is no more than classification, which many artificial neural networks can perform. Classification represents a reduction of superfluous information that can be useful and economic for storage if the symbol retains the important attributes of the referent. If the output of a classifier itself serves as input to another level, it can simplify the process for learning higher-order relationships and patterns. The intuitive idea is that a simple, condensed representation, being a shorthand version of a more complex representation, can be manipulated more easily – and needs less storage capacity – than the complete representation of the item. We can more easily memorize items if we attribute symbols (e.g., words) to them, so it might well be that neural representations of symbols provide such a reduced representation. We will use the term symbolic representation in this way.

However, the manipulation of symbols requires something more than just bottom-up processing and classification of input patterns; it requires two-way interactions. How can such a shorthand version be represented in a neural network in such a way that manipulation is possible? One proposal is that, in addition to the complete, distributed memory trace describing the different attributes of the item (e.g., leaves, branches and roots as attributes of trees), special units exist that are connected to the feature detectors for the attributes. The presence of a particular pattern is signaled not only implicitly by the overall pattern of activity but also explicitly by the presence or absence of activity in a smaller number of specialist neurons connected to the attributes. The activities in these neurons form a shorthand or symbolic representation. These connections correspond to the K-lines that Minsky (1986) proposed for a feedforward net. However, to make a procedure for manipulation possible, the connections must also be recurrent: activity in the symbol units must activate units associated with the attributes. Models of this kind have been proposed for language production and comprehension (e.g., McClelland & Rumelhart 1981; McClelland & Kawamoto 1986; MacKay 1987); the networks typically have an explicit hierarchy corresponding to levels of abstraction, that is, phoneme, word and semantics. Considering all the units to be part of one large recurrent net, one could say that in the absence of prior activity in the symbol units, the network has an attractor that includes activation of some symbol units. If, however, the units representing one symbol are strongly interconnected and, if only these are activated, the attractor formed by the corresponding global pattern could be stabilized, allowing it to be established faster or from a wider input region.

If, in addition, relationships among the symbol units can be learned, one could imagine that, because fewer connections are necessary to connect a few "symbol" units, these more abstract representations of an item permit more associations to be made. One could further speculate that this smaller number of connections could provide easier links to a larger domain of stored information.

Such a scheme can also explain how stored information can be manipulated, that is, applied in a new context other than in the procedure in which it was learned. Take the following example of how this might work. You need to turn a screw, but no screwdriver is available. The task "turn a screw" is represented by an input vector. Because the task has been solved often, there is a deep, primary attractor representing "use a screwdriver". Because the sensory input says "there is no screwdriver", this primary attractor, or perhaps only its most active units, are inhibited, forcing the system to search for another attractor that might fit the task. It is clear that in the naive condition the search for other suitable solutions cannot be made on the symbolic level, that is, the intelligence demonstrated in the behavior of finding a novel alternative to a screwdriver cannot rest on already having a list of alternatives. It could arise through intelligence of process if the activation of the screwdriver assembly sufficiently excites relevant attributes on a lower level so this activity moves the system to an alternative attractor, one influenced by relevant sensory inputs (one happens to be looking at some object with a thin edge) or by symbol units engaging similar attributes (e.g., a coin with a thin edge) or by their combination. As a result, the sensory input "coin" may elicit an attractor combining coin and task. This alternative will be tried and, if successful, a new link (turn screw → use a coin) could become established, that is, an attractor involving coin and screwdriver or coin and the task of turning a screw may begin to form. The next time a screw must be turned and no screwdriver is at hand, this link may be sufficiently established to find the alternative faster and more directly than by search via shared attributes.

This example also illustrates two further points. First, there is a need for a module that can act upon and modulate the activity in the network by inhibiting some units and allowing excitation in others to grow. If the attractor for screwdriver is excited but no screwdriver is at hand, this additional module must let activity continue to grow in other attractors. This aspect is cognitive in that it in a sense evaluates the possible realization of one proposed solution, that encoded in the current state of activity. Searching for a realizable attractor requires time. This procedure would correspond to the manipulation of knowledge and to "thinking"; it is different from the application of existing procedural knowledge. Second, the speed and directedness of this search procedure can show more or less intelligence of process. An "intelligent" system should concentrate on "promising" regions within its distributed memory (the frame problem: Lanz 1994). The better it can do this, the more intelligent it would be judged to be.

Several artificial neural networks represent steps toward the mental manipulations described here as minimal cognition (e.g., Dyna: Sutton 1991; the ART models Grossberg 1988; and Carpenter & Grossberg 1993). Such systems are able to generate new categories and actions when confronted with a new stimulus that is not sufficiently close to previously stored patterns. Subsequently, the now familiar stimulus can be treated procedurally. In part, these searches follow the procedures for associative storage searches as described for typical neuronal nets. Hence, they, too, can be considered to be prerational.

However, during thinking, humans also seem to apply logical rules. What mechanisms might be necessary for the brain to apply logical rules? At first sight, this might seem trivial because some familiar simple circuits carry out logical operations (e.g., AND or XOR circuits). A module that simply performs the given operation on its inputs shows logic of process as a result of design, analogous to the distinction made with respect to intelligence. Alternatively, logical operations could be stored more abstractly, like symbols, so that several alternative operations could be applied to the same data. Superficially, this would provide logic of content, but it could be realized in a similar manner to that described above for manipulation of connections involving symbol units. Where do the rules of logic come from? This question has a long history in philosophical discussions (e.g., Putnam 1971). For a biologist, it is tempting to assume that they represent abstractions of regularities in the natural world that are learnt during ontogeny or encoded genetically during evolution. An individual can acquire such a natural law either as declarative knowledge or as a procedure, and either consciously or unconsciously. In modern philosophy, such an empiricist view, like those propounded by Mill and Hume, is generally rejected by more traditional philosophers of logic, but Goodman (1973) proposed a similar, pragmatic view of logical and empirical laws as the product of a continuous comparison of individual experiences and general laws.

Up to now we have avoided terms referring to the introspective realm, that is, conscious knowledge or subjective experience. Some information from the procedural domain clearly is not accessible to conscious awareness. This is true not only in rather obvious cases like low-level reflexes but also at unexpectedly high levels (e.g., blindsight, see Bridgeman et al., this volume). There are, however, non-declarative phenomena (states) that can enter consciousness, such as feelings or smells. By definition, all declarative knowledge can become conscious. One might assume that the application of logical rules is constrained to the conscious domain, but this can be questioned; a logic circuit carrying out an XOR operation obviously has no consciousness. Thus, one might distinguish between the consciously penetrable domain and the consciously impenetrable domain.

In any event, there are connections in both directions between these domains. As mentioned above, declarative knowledge about skills can be transferred through practice to the procedural domain. In certain conditions, activity in the conscious domain interferes with activity in the procedural domain and downgrades procedural performance (see Bridgeman et al., this volume).

A final set of questions that remain completely open concerns the internal aspect or subjective experience. The internal state of a human being can be determined, in theory, by objective, externally measurable parameters like the state of all the neurons, the internal milieu with its hormones and modulators, and the physical body; but the internal state is experienced by the human internally or subjectively. We assume that this is also the case for many animals. The question is whether the existence of this internal aspect is essential in a causal sense. In other words, does the phenomenon of internal experience enable the system to behave more intelligently? To avoid misunderstanding, we do not mean to question the necessity of the processes that provide the physiological basis for our introspective experience. These most probably are functionally important. The question concerns only the fact of the internal experience as such. Conceivably, it is a necessary consequence of the unification of domains for the global optimization that biologists refer to as fitness, just as pleasure can be seen as the common currency for selecting among actions in different realms. Thus, the open questions include the following. What conditions have to be fulfilled for the phenomenon of internal aspect to occur? Can we understand how this phenomenon arises in humans? Could it also occur in artificial systems? Is the internal aspect necessary for the performance of the system; that is, could all the processes affecting observable behavior be implemented without any internal aspect? Konrad Lorenz (1963) posed some of these same questions with respect to subjective experience in animals and concluded that many are, in principle, unanswerable.

In conclusion, this review of intelligence in artificial and natural systems suggests that the borderline between prerational and rational intelligence disappears in the sense that declarative knowledge seems to be a special form of procedural knowledge, and rational intelligence a special form of prerational mechanisms, just as the border between procedural and declarative knowledge in biological systems would disappear if we understood the physiological procedures underlying declarative knowledge. This view is not altered by the evidence that differences in the structure and organization of the two domains may lend themselves to processing with different combinations of architectures and elementary properties, just as the activation properties of output units in artificial neural networks can be adjusted for continuous or discrete transformations. Therefore, the anatomical separation and differences in temporal plasticity are not surprising. However, there is every reason to believe that the basic elemental and network properties are similar, or in other words, declarative knowledge and logical thought are just types of procedures implemented in the prerational machinery of the nervous system. The review of artificial systems shows that we are still very far from being able to match the overall performance and adaptive behavior of animals, although results in limited domains are more nearly comparable.

However, both these artificial systems and the increasing knowledge about biological systems already suggest in broad detail how intelligence arises in biological systems and point out directions for achieving artificial systems with similar capabilities. The road to travel is still long and there may be surprises, but the conceptual framework appears to be available. One question of continuing interest in comparing natural and artificial systems concerns the necessity of biological organization and function for the achievement of equivalent performance. Animals are limited by their evolutionary history. Some features may simply reflect this history, leaving open the possibility for alternative systems with significantly different structures. Details of modularity and connectivity in the brain may be one example of specific, historically conditioned solutions; general principles of modularity, plasticity and dynamics may be more universal. Still more intriguing is the question of whether the internal aspect is another such necessity or whether it would necessarily arise in artificial systems with behavior of equal complexity and intelligence. As simulation and biology approach one another, it may be possible to reply to some of the questions Lorenz (1963) considered unanswerable in principle.

*Cleveland State University, Cleveland, Ohio, USA
**Université Libre de Bruxelles, Brussels, Belgium
***Universität Bielefeld, Germany

REFERENCES

Baldi, P., & W. Heiligenberg (1988). How sensory maps could enhance resolution through ordered arrangements of broadly tuned receivers. *Biological Cybernetics* **59**, 313–318.

Barlow, H.B. (1972). Single units and cognition: a neurone doctrine for perceptual psychology. *Perception* **1**, 371–394.

Barto, A.G., R.S. Sutton, & C.W. Anderson (1983). Neuronlike adaptive elements that can solve difficult learning control problems. *IEEE Transactions Systems, Man and Cybernetics* **13**, 834–846.

Beer, R.D., & J.C. Gallagher (1992). Evolving dynamical neural networks for adaptive behavior. *Adaptive Behavior* **1**, 91–122.

Belardetti, F., & S.A. Siegelbaum (1988). Up- and down-modulation of single $K+$ channel function by distinct second messengers. *Trends in Neurosciences* **11**, 235–238-

Bell, C.C. (1982). Properties of a modifiable efference copy in an electric fish. *Journal of Neurophysiology* **47**, 1043–1056.

Bell, C.C. (1984). Effects of motor commands on sensory inflow, with examples from electric fish. In L. Bolis, R.D. Keynes, & S.H.P. Maddrell (eds.), *Comparative physiology of sensory systems* (pp. 637–646). Cambridge, MA: Cambridge University Press.

Bergold, P.J., J.D. Sweatt, I. Winicov, K.R. Weiss, E.R. Kandel, & J.H. Schwartz (1990). Protein synthesis during acquisition of long-term facilitation is needed for the persistent loss of regulatory subunits of the *Aplysia* cAMP-dependent protein kinase. *Proceedings of the National Academy of Science USA* **87**, 3788–3791

Braitenberg, V. (1984). *Vehicles. Experiments in synthetic psychology.* Cambridge, MA: MIT Press.

Brennan, P., H. Kaba, & F.B. Keverne (1990). Olfactory recognition: a simple memory system. *Science* **250**, 1223–1226.

Brooks, R.A. (1986). A robust layered control system for a mobile robot. *IEEE Journal of Robotics and Automation* **RA-2**, 14–23.

Brooks, R.A. (1991). Intelligence without reason. *Proceedings of the Twelfth International Joint Conference on Artificial Intelligence, IJCAI-91* (pp. 569–595). San Mateo, CA: Morgan Kaufman.

Brooks, R.A. (1991). Challenges for complete creature architectures. In J.-A. Meyer & S. Wilson (eds.), *From animals to animats* (pp. 434–443). Cambridge, MA: MIT Press.

Brooks, V.B. (1986). *The neural basis of motor control.* Oxford, UK: Oxford University Press.

Brunn, D.E., & J. Dean (1994). Intersegmental and local interneurons in the metathorax of the stick insect *Carausius morosus* which monitor middle leg position. *Journal of Neurophysiology* **72**, 1208–1219.

Bullock, T.H. (1979). Evolving concepts of local integrative operations in neurons. In F.O. Schmitt & F.G. Worden (eds.), *The neurosciences fourth study program* (pp. 52–60). Cambridge, MA: MIT Press.

Bullock, T.H., & G.A. Horridge (1965). *Structure and function in the nervous system of invertebrates.* San Francisco: W.H. Freeman.

Byrne , J.H., R. Zwartjes, R. Homayouni, S.D. Critz, & A. Eskin (1993). Roles of second messenger pathways in neuronal plasticity and in learning and memory. In S. Shenolikar & A.C. Nairn (eds.), *Model systems in signal transduction* (Advances in Second Messenger and Phosphoprotein Research Series, Vol. 27) (pp. 47–108). New York: Raven Press.

Carew, T.J., R.D. Hawkins, T.W. Abrams, & E.R. Kandel (1984). A test of Hebb's postulate at identified synapses which mediate classical conditioning in *Aplysia*. *Journal of Neuroscience* **4**, 1217–1224.

Cabanac, M. (1991). Pleasure: the answer to conflicting motivations. In J.-A. Meyer & S. Wilson (eds.), *From animals to animats* (pp. 206–212). Cambridge, MA: MIT Press.

Carpenter, G.A., & S. Grossberg (1993). Normal and amnesic learning, recognition and memory by a neural model of cortico-hippocampal interactions. *Trends in Neurosciences* **16**, 131–137.

Cruse, H. (1990.) What mechanisms coordinate leg movement in walking arthropods? *Trends in Neurosciences* **13**, 15–21.

Cruse, H., J. Dean, H. Heuer, & R.A. Schmidt (1990). Utilization of sensory information for motor control. In O. Neumann & W. Prinz (eds.), *Relationships between perception and action: Current approaches* (pp. 43–79). Berlin: Springer-Verlag.

Cruse, H., D. Brunn, Ch. Bartling, J. Dean, M. Dreifert, T. Kindermann, & J. Schmitz (1995). Walking: A complex behavior controlled by simple networks. *Adaptive Behavior* **3**, 385–417.

Damasio, A.R. (1989). Mechanisms of face recognition. In A.W. Young & H.D. Ellis (eds.), *Handbook of research on face processing* (pp. 405–425). New York: North Holland.

Darwin, C. (1859). *The origin of species by means of natural selection.* London: John Murray.

Davis, H.P., & L.R. Squire (1984). Protein synthesis and memory: a review. *Physiological Bulletin* **96**, 518–559.

Dean, J. (1989). Leg coordination in the stick insect, *Carausius morosus*: Effects of cutting thoracic connectives. *Journal of Experimental Biology* **145**, 103–131.

Dean, J. (1990.) Coding proprioceptive information to control movement to a target: Simulation with a simple neural network. *Biological Cybernetics* **63**, 115–120.

Dean, J. (1991). A model of leg coordination in the stick insect, *Carausius morosus*. II. Description of the kinematic model and simulation of normal step patterns. *Biological Cybernetics* **64**, 403–411.

Dean, J. (1998). Animats and what they can tell us. *Trends in Cognitive Sciences* **1**, 60–67.

Delcomyn, F. (1980). Neural basis of rhythmic behavior in animals. *Science* **210**, 492–498.

Deneubourg, J.L., S. Aron, S. Goss, J.M. Pasteels, & G. Duerink (1986). Random behavior, amplification processes and number of participants: How they contribute to the foraging properties of ants. Proceedings of the Los Alamos Conference on Games. *Learning and Evolution, Physica* **22 D**, 176–186.

Deneubourg J.L., S. Goss, J.M. Pasteels, D. Fresnau, & J.P. Lachaud (1987). Self-organizing mechanisms in ant societies (II): Learning in foraging and division of labour. In J.M. Pasteels & J.L. Deneubourg (eds.), *From individual to collective behaviour in social insects. Experientia suppl. 54* (pp. 177–196). Basel, Switzerland: Birkhäuser.

Deneubourg J.L., J.M. Pasteels, & J.C. Verhaege (1984). Quand l'erreur alimente l'imagination d'une société: Le cas des fourmis. *Nouvelles de la Science et des Technologies* **2**, 47–52.

Eckhorn, R., R. Bauer, W. Jordan, M. Brosch, W. Kruse, M. Munk, & H.J. Reitboeck (1988). Coherent oscillations: A mechanism for feature linking in the visual cortex? *Biological Cybernetics* **60**, 121–130.

Egelhaaf, M. (1987). Dynamic properties of two control systems underlying visually guided turning in house-flies. *Journal of Comparative Physiology* A **161**, 777–783.

Egelhaaf, M. (1990). Spatial interactions in the fly visual system leading to selectivity for small-field motion. *Naturwissenschaften* **77**, 182–185.

Eibl-Eibesfeldt, I. (1980). Jumping on the sociobiology bandwagon. *Behavioral Brain Sciences* **3**, 631–634.

Engel, A.K., P. König, A.K. Kreiter, & W. Singer (1991). Interhemispheric synchronization of oscillatory neuronal responses in cat visual cortex. *Science* **252**, 1177–1179.

Fabré, J.H. (1989). *Wunder des Lebendigen*. Zürich: Artemis Verlag. Translated from *Souvenirs Entomologiques*.

Favilla, M., W. Hening, & C. Ghez (1989). Trajectory control in targeted force impulses. VI. Independent specification of response amplitude and direction. *Experimental Brain Research* **75**, 280–294.

Felleman, D.J., & D.G. Van Essen (1991). Distributed hierarchical processing in the primate cerebral cortex. *Cerebral Cortex* **1**, 1–47.

Franceschini, N., Pichon, J.M., & Blanes, C. (1992). From insect vision to robot vision. *Philosophical Transactions of the Royal Society London* **B 337**, 283–294.

Gallagher, J.C., & R.D. Beer (1993). A qualitative dynamical analysis of evolved locomotion controllers. In J.-A. Meyer, H.L. Roitblat, & S.W. Wilson (eds.), *From animals to animats 2* (pp. 71–80). Cambridge, MA: MIT Press.

Garcia, J., D.J. Kimmeldorf, & R.A. Koelling (1955). Conditioned aversion to saccharin resulting from exposure to gamma radiation. *Science* **122**, 157–158.

Georgopoulos, A.P., M. Taira, & A. Lukashin (1993). Cognitive neurophysiology of the motor cortex. *Science* **260**, 47–52.

Ghez, C., W. Hening, & M. Favilla (1989). Parallel interacting channels in the initiation and specification of motor response features. In M. Jeannerod (ed.), *Attention and performance, XIII* (pp. 265–293). Hillsdale, NJ: Lawrence Erlbaum Assoc.

Ghez, C., J. Gordon, M.F. Ghilardi, C.N. Christakos, & S.E. Cooper (1990). Roles of proprioceptive input in the programming of arm trajectories. *Cold Spring Harbor Symposium on Quantitative Biology* **LV** (pp. 837–847).

Ghirardi, M., O. Braha, B. Hochner, P.G. Montarolo, E.R. Kandel, & N. Dale (1992). Roles of PKA and PKC in facilitation of evoked and spontaneous transmitter release at depressed and nondepressed synapses in *Aplysia* sensory neurons. *Neuron* **9**, 479–489.

Glansdorff, P., & I. Prigogine (1971). *Thermodynamic theory of structure, stability and fluctuations*. London: Wiley.

Goldsmith, B.A., & T.W. Abrams (1991). cAMP modulates multiple $K+$ currents, increasing spike duration and excitability in *Aplysia* sensory neurons. *Proceedings of the National Academy of Science USA* **89**, 11481–11485.

Goodman, N. (1973). *Fact, fiction, and forecast*. Indianapolis, IN: Bobbs-Merrill.

Grossberg, S. (1988). *Neural networks and natural intelligence*. Cambridge, MA: MIT Press.

Hamilton, W.D. (1964). The genetical theory of social behaviour. I and II. *Journal of Theoretical Biology* **7**, 1–52.

Hammer, M., & R. Menzel (1995). Learning and memory in the honeybee. *Journal of Neuroscience* **15**, 1617–1630.

Hartmann, G. (1992). Motion induced transformations of spatial representations: Mapping 3-D information onto 2-D. *Neural Networks* **5**, 823–834.

Hartmann, G., & R. Wehner (1995). The ant's path integration system: A neural architecture. *Biological Cybernetics* **73**, 483–497.

Hawkins, R.D., T.W. Abrams, T.J. Carew, & E.R. Kandel (1983). A cellular mechanism of classical conditioning in *Aplysia*: Activity-dependent amplification of presynaptic facilitation. *Science* **219**, 400–405.

Hawkins, R.D., E.R. Kandel, & S.A. Siegelbaum (1993). Learning to modulate transmitter release: Themes and variations in synaptic plasticity. *Annual Review of Neuroscience* **16**, 625–665.

Hebb, D.O. (1949). *The organization of behavior*. New York: Wiley.

Hertz, J.A., R.G. Palmer, & A.S. Krogh (1991). *Introduction to the theory of neural computation*. Redwood City, CA: Addison-Wesley.

Hodos, W. (1982). Some perspectives on the evolution of intelligence and the brain. In D.R. Griffin (ed.), *Animal mind – Human mind* (pp. 33–55). Berlin: Springer Verlag.

Holst, E. von (1950). Die Arbeitsweise des Statolithenapparates bei Fischen. *Zeitschrift für vergleichende Physiologie* **32**, 60–120.

Holst, E. von (1969). *Zur Verhaltensphysiologie bei Tieren und Menschen*. München: R. Piper & Co.

Holst, E. von, & H. Mittelstaedt (1950). Das Reafferenzprinzip. *Naturwissenschaften* **37**, 464–476.

Hooper, S.L., & M. Moulins (1990). Cellular and synaptic mechanisms responsible for a long-lasting restructuring of the lobster pyloric network. *Journal of Neurophysiology* **64**, 1574–1589.

Houston, J.P. (1991). *Fundamentals of learning and memory*. 4th ed. Fort Worth, TX: Harcourt, Brace, Jovanovich.

Hubel, D.H., T.N. Wiesel, & S. LeVay (1977). Plasticity of ocular dominance columns in monkey striate cortex. *Philosophical Transactions of the Royal Society London* **278**, 377–409.

Johansson, G. (1973). Visual perception of biological motion and a model for its analysis. *Perception and Psychophysics* **14**, 201–211.

Kandel, E.R., J.H. Schwartz, & M. Jessel (eds.), (1991). *Principles of neural science*. 3rd. Amsterdam: Elsevier.

Kendrick, M.K., F. Lévy, & F.B. Keverne (1992). Changes in the sensory processing of olfactory signals induced by birth in sheep. *Science* **256**, 833–836.

Kolb, B., & I.Q. Whishaw (1990). *Fundamentals of neuropsychology*. 3rd ed. New York: W.H. Freeman.

Krebs, J.R., & N.B. Davies (eds.), (1978). *Behavioural ecology: An evolutionary approach*. Oxford, UK: Blackwell.

Krotkov, E., J. Bares, T. Kanade, T. Mitchell, R. Simmons, & R. Whittaker (1991). Ambler: A six-legged planetary rover. *1991 ICAR. Proceedings of the Fifth International Conference on Advanced Robotics*, Vol. 2, 717–722.

Kupfermann, I. (1991). Learning and memory. In E.R. Kandel, J.H. Schwartz, & T.M. Jessel (eds.), *Principles of neural science*. 3rd. (pp. 997–1008). Amsterdam: Elsevier.

Kupfermann, I., & K.R. Weiss (1978). The command neuron concept. *Behavioral Brain Sciences* **1**, 3–39.

Lashley, K.S. (1950). In search of the engram: Physiological mechanisms in animal behaviour. In J.F. Danielli & R. Brown (eds.), *Symposium of the Society for Experimental Biology, vol. 4,* (pp. 454–482). Cambridge, UK: Cambridge University Press.

Lehky, S.R., & T.J. Sejnowski (1990). Neural model of stereoacuity and depth interpolation based on distributed representation of stereo disparity. *Journal of Neuroscience* **10**, 2281–2299.

Llinás, R.R. (1988). The intrinsic electrophysiological properties of mammalian neurons: Insights into central nervous system function. *Science* **242**, 1654–1664.

Llinás, R.R. (1990). Intrinsic electrical properties of nerve cells and their role in network oscillation. In: *The brain. Cold Spring Harbor Symposium on Quantitative Biology* **LV**, 933–938.

Lockery, S.R., & W.B. Kristan (1990). Distributed processing of sensory information in the leech. II. Identification of interneurons contributing to the local bending reflex. *Journal of Neuroscience* **10**, 1816–1829.

Lockery, S.R., & T.J. Sejnowski (1993). A lower bound on the detectability of nonassociative learning in the local bending reflex of the medicinal leech. *Behavioral Neural Biology* **59**, 208–224.

Lorenz, K. (1950). The comparative method in studying innate behaviour patterns. In: *Physiological mechanisms in animal behavour. Symposium of the Society for Experimental Biology* **4**, 221–268.

Lorenz, K. (1963). Haben Tiere ein subjektives Erleben? Reprinted in: *Über tierisches und menschliches Verhalten. Aus dem Werdegang der Verhaltenslehre. Gesammelte Abhandlungen Band II* (pp. 359–374). München: Piper.

Lorenz, K. (1965). *Evolution and modification of behavior*. Chicago, ILL: University of Chicago Press.

Lytton, W.W., & T.J. Sejnowski (1991). Simulations of cortical pyramidal neurons synchronized by inhibitory interneurons. *Journal of Neurophysiology* **66**, 1059–1079.

MacKay, D.G. (1987). *The organization of perception and action: A theory for language and other cognitive skills*. New York: Springer.

Maes, P. (1991). A bottom-up mechanism for behavior selection in an artificial creature. In J.-A. Meyer & S. Wilson (eds.,) *From animals to animats* (pp. 238–246). Cambridge, MA: MIT Press.

Maes, P. (1993). Behavior-based artificial intelligence. In J.-A. Meyer, H.L. Roit-blat, & S.W. Wilson (eds.), *From animals to animats 2* (pp. 2–10). Cambridge, MA: MIT Press.

Matarić, M.J. (1992). Designing emergent behaviors: from local interactions to collective intelligence. In J.-A. Meyer, H.L. Roitblat, & S.W. Wilson (eds.), *From animals to animats 2* (pp. 432–441). Cambridge, MA: MIT Press.

McClelland, J.L., & D.E. Rumelhart (1981). An interactive activation model of context effect in letter perception: Part 1. An account of basic findings. *Psychological Review* **88**, 375–407.

McClelland, J.L., & A.H. Kawamoto (1986). Mechanisms of sentence processing: Assigning roles to constituents. In J.L. McClelland, D.E. Rumelhart, & the PDP Research Group (eds.), *Parallel distributed processing: Explorations in the microstructure of cognition. Vol 2: Psychological and biological models* (pp. 272–325). Cambridge, MA: MIT Press.

McCullock, W.S., & Pitts, W. (1943). A logical calculus of ideas immanent in nervous activity. *Bulletin Mathematical Biophysics* **5**, 115–133.

McFarland, D. (1985). *Animal behaviour*. London: Pittman.

McFarland, D. (1991). Defining motivation and cognition in animals. *International Studies in the Philosophy of Science* **5**, 153–170.

McFarland, D. (2000). Rational behaviour of animals and machines. In H. Cruse, H. Ritter, & J. Dean (2000), *Prerational intelligence: Adaptive behavior and intelligent systems without symbols and logic, Vol. 1* (pp. 31–42). Dordrecht, The Netherlands: Kluwer Academic Publishers.

McFarland, D., & A. Houston (1981). *Quantitative ethology: The state-space approach*. London: Pitman Books.

McFarland, D., & T. Bösser (1993). *Intelligent behavior in animals and robots*. Cambridge, MA: MIT Press.

McMahon, T.A. (1984). *Muscles, reflexes, and locomotion*. Princeton, NJ: Princeton University Press.

Merzenich, M.M., & J.H. Kaas (1982). Reorganization of mammalian somatosensory cortex following peripheral nerve injury. *Trends in Neurosciences* **5**, 434–436.

Miller, G.A., E. Galanter, & K.H. Pribram (1960). *Plans and the structure of behavior*. New York: Holt, Rinehart and Winston.

Minsky, M.L. (1986). *The society of mind*. New York: Simon & Schuster.

Minsky, M.L., & S.A. Papert (1969). *Perceptrons*. Cambridge, MA: MIT Press

Mittelstaedt, M. (1964). Basic control patterns of orientational homeostasis. *Symposium of the Society for experimental Biology* **18**, 365–385.

Möhl, B. (1989) 'Biological noise' and plasticity of sensorimotor pathways in the locust flight system. *Journal of Comparative Physiology* **A, 166**, 75–82.

Müller-Wilm, U., J. Dean, H. Cruse, H.J. Weidemann, J. Eltze, & F. Pfeiffer (1992). Kinematic model of a stick insect as an example of a 6-legged walking system. *Adaptive Behavior* **1**, 155–169.

Newell, A. (1990) *Unified theories of cognition.* Cambridge, MA: Harvard University Press.

Newell, A., & H.A. Simon (1972). *Human problem solving.* Engelwood Cliffs, NJ: Prentice-Hall.

Nicolis, G., & I. Prigogine (1988). *Exploring complexity.* New York: Freeman & Co.

Olshausen, B.A., C.H. Anderson, & D.G. Van Essen (1993). A neurobiological model of visual attention and invariant pattern recognition based on dynamic routing of information. *Journal of Neuroscience* **13**, 4700–4719.

Oram, M., & D. Perrett (1992). Time course of neural responses discriminating different views of the face and head. *Journal of Neurophysiolgy* **68**, 70–84.

Oster, G.F., & E.O. Wilson (1978). *Caste and ecology in the social insects.* Princeton, NJ: Princeton University Press.

Pasteels, J.M., J.L. Deneubourg, & S. Goss (1987a). Transmission and amplification of information in a changing environment: The case of insect societies. In I. Prigogine & M. Sanglier (eds.), *Laws of nature and human conduct: Specificities and unifying themes* (pp. 129–156). Brussels: GORDES, Brussels.

Pasteels, J.M., J.L. Deneubourg, & S. Goss (1987b). Self-organization mechanisms in ant societies (I): Trail recruitment to newly discovered food sources. In J.M. Pasteels & J.-L. Deneubourg (eds.), *From individual to collective behavior in social insects* (pp. 155–175). Basel: Birkhäuser.

Perret, D.I., A.J. Mistlin, & A.J. Chitty (1987). Visual neurones responsive to faces. *Trends in Neurosciences* **10**, 358–364.

Pfeiffer, F., H.J. Weidemann, & P. Danowski (1990). Dynamics of the walking stick insect. *Proceedings of the 1990 IEEE International Conference on Robotics and Automation* (pp. 1458–1463).

Piaget, J., & B. Inhelder (1977). *Von der Logik des Kindes zur Logik des Heranwachsenden.* Olten, Switzerland: Walter-Verlag.

Pringle, J.W.S. (1957). *Insect flight.* Cambridge, UK: Cambridge University Press.

Putnam, H. (1971). *Philosophy of logic.* London: Allen Unwin.

Reichert, H., & C.H.F. Rowell (1986). Neuronal circuits controlling flight in the locust: How sensory information is processed for motor control. *Trends in Neurosciences* **9**, 281–283.

Revusky, S.H., & J. Garcia (1970). Learned associations over long delays. In G.H. Bower (ed.), *Psychology of learning and motivation*, Vol. 4 (pp. 1–83). New York: Academic Press.

Roberts, A., & B.M.H. Bush (eds.), (1981. *Neurones without impulses: Their significance for vertebrate and invertebrate nervous systems.* Cambridge, UK: Cambridge University Press.

Rolls, E.T. (1991). Neural organization of higher visual functions. *Current Opinion in Neurobiology* **1**, 274–278.

Rumelhart, D.E., G.E. Hinton, & R.J. Williams (1986). Learning internal representations by error propagation. In D.E. Rumelhart, J.L. McClelland, & the PDP
Research Group (eds.), *Parallel distributed processing: Explorations in the microstructure of cognition, Vol 1: Foundations* (pp. 318–362). Cambridge, MA:
MIT Press.

Ryle, G. (1949). *The concept of mind.* London: Hutchinson.

Seligman, M.E.P. (1970). On the generality of the laws of learning. *Psychological
Review* 77, 406–418.

Seligman, M.E.P., & J. Hager (eds.), (1972). *Biological boundaries of learning.*
New York: Appleton.

Shepherd, G.M. (ed.), (1990). *The synaptic organization of the brain.* 3rd ed. Oxford, UK: Oxford University Press.

Sherrington, C.S. (1941). *Man on his nature.* Cambridge, UK: Cambridge University Press.

Singer, W. (1993). Synchronization of cortical activity and its putative role in information processing and learning. *Annual Review of Physiology* 55, 349–374.

Smithers, T. (1992). Taking eliminative materialism seriously: A methodology for
autonomous systems research. In F.J. Varela & P. Bourgine (eds.), *Toward a
practice of autonomous systems. Proceedings of the first european conference
on artificial life* (pp. 31–40). Cambridge, MA: MIT Press.

Softly, W.R., & C. Koch (1993). The highly irregular firing of cortical cells is in-
consistent with temporal integration of random EPSPs. *Journal of Neuroscience*
13, 334–350.

Sperry, R.W. (1950). Neural basis of the spontaneous optokinetic response produced
by visual inversion. *Journal of Comparative Physiology and Psychology* 43,
482–489.

Steels, L. (1991). Towards a theory of emergent functionality. In J.-A. Meyer &
S.W. Wilson (eds.), *From animals to animats* (pp. 451–461). Cambridge, MA:
MIT Press.

Stephens, D.W., & J.R. Krebs (1986). *Foraging theory.* Princeton, NJ: Princeton
University Press.

Suga, N., J. Yan, & Y. Zhang (1997). Cortical maps for hearing and egocentric
selection for self-organization. *Trends in Cognitive Sciences* 1, 13–20.

Sutton, R.S. (1991). Reinforcement learning architectures for animats. In J.-A.
Meyer & S.W. Wilson (eds.), *From animals to animats* (pp. 288–296). Cambridge, MA: MIT Press.

Theraulaz, G., S. Goss, J. Gervet, & J.-L. Deneubourg (1991). Task differentiation
in Polistes wasp colonies: A model for self-organizing groups of robots. In
J.-A. Meyer & S.W. Wilson (eds.), *From animals to animats* (pp. 346–355).
Cambridge, MA: MIT Press.

Tulving, E. (1987). Multiple memory systems and consciousness. *Human Neurobiology* 6, 67–80.

Turvey, M.T. (1977). Preliminaries to a theory of action with reference to vision. In R. Shaw & J. Bransford (eds.), *Perceiving, acting and knowing: Toward an ecological psychology* (pp. 211–265). Hillsdale, NJ: Lawrence Erlbaum Ass.

Ullman, S. (1979). *The interpretation of visual motion*. Cambridge, MA: MIT Press.

Van Essen, D.C., D.J. Felleman, E.A. DeVoe, J. Olavarria, & J. Knierim (1990). Modular and hierarchical organization of extrastriate visual cortex in the macaque monkey. *The brain. Cold Spring Harbor Symposium on Quantitative Biology* **LV** (pp. 679–696).

Verschure, P.F.M.J., B.J.A. Kröse, & R. Pfeifer, R. (1992). Distributed adaptive control: The self-organization of structured behavior. *Robotics and Autonomous Systems* **9**, 181–196.

Walters, E.T., & J.H. Byrne (1983). Associative conditioning of single sensory neurons suggests a cellular mechanism for learning. *Science* **219**, 405–408.

Webb, B. (1994). Robotic experiments in cricket phonotaxis. In D. Cliff, P. Husbands, J.-A. Meyer, & S.W. Wilson (eds.), *From animals to animats 3: Proceedings of the Third International Conference on the Simulation of Adaptive Behaviour* (pp. 45–54). Cambridge, MA: MIT Press.

Wendler, G. (1985). Insect locomotory systems: Control by proprioceptive and exteroceptive inputs. In M. Gewecke & G. Wendler (eds.), *Insect locomotion* (pp. 245–254). Hamburg: Parey

Wilson, E.O. (1971). *The insect societies*. Cambridge, MA: Belknap Press.

Wilson E.O. (1975. *Sociobiology. The new synthesis*. Cambridge, MA: Belknap Press.

Wilson, E.O., & B. Hölldobler (1988). Dense hierarchies and mass comunication as the basis of organization in ant colonies. *Trends in Ecology and Evolution* **3**, 65–67.

Yovell, Y., & T.W. Abrams (1992). Temporal asymmetry in activation of an adenylyl cyclase by calcium and transmitter may explain temporal requirements of conditioning. *Proceedings of the National Academy of Science, USA*, **89**, 6526–6530.

BRUCE BRIDGEMAN*, GUY CELLERIER**, JACQUES PAILLARD***,
and BORIS VELICHKOVSKY****

PRERATIONAL INTELLIGENCE FROM THE PERSPECTIVE OF PSYCHOLOGY

1. INTRODUCTION

The current influence of the "behavior-based" approach to robotics (Brooks 1991), as well as the notion of "prerational intelligence" introduced in the present volume, challenge psychologists in one of their traditional fields of inquiry.

Historically, an anthropocentric, mentalistic view of intelligence has dominated the field of psychology. Rationality is generally viewed as the production of "value-driven" mental operations that manipulate stored symbolic knowledge according to logical rules. In psychology, symbolic reasoning is generally considered to be closely tied to language and therefore specifically human. "Irrational" behaviors studied by human psychiatry and neuropsychology arise from disorders of these operations and do not have a definite extension to animal behavior.

Moreover, owing to its basic "value-dependency" the concept of rationality remains controversial in psychology in the context of defining "intelligent" thinking or behavior in general. The usefulness of this notion from this perspective will not be further discussed here. More pragmatically we may question and evaluate the relevance and originality of the approach suggested by animal research (Brooks 1991) from the perspective of psychology.

The modern theory of information processing, the pervading influence in contemporary psychology, uses a computational theory of mind based primarily on the manipulation of symbolic representations. Until recently this was taken as the only possible formal access to high-level cognition. This explains modern cognitivism's rather close connection with classical artificial intelligence (AI) and its progressive loss of grounding in underlying neurobiological mechanisms.

In contrast, the new "behavior-based robotics" focuses on low-level basic adaptive performances in animals or simple robots in real-world environments. It assumes that studying these performances will reveal some kind of "prerational" lower level expression of intelligent behavior (McFarland & Bösser 1993) which in turn could provide a radically new understanding

J. Dean et al. (eds.), Prerational Intelligence: Interdisciplinary Perspectives on the Behavior of Natural and Artificial Systems, 89–134.

of high-level phenomena such as communication, representation, intentional action and even language (Steward 1994). Behavior-based robotics is, therefore, presented mainly as a potential alternative to the classical AI approach.

The study of basic adaptive mechanisms and elementary functions in living organisms is the traditional bottom-up strategy of biology. The difficulty encountered by pure reflexologic or behavioristic approaches to bridging the gap to psychiatry's and cognitive psychology's top-down study of high-order mental processes is historically established. Thus, a psychologist would question the degree to which the production of such elementary adaptive processes can be described as "intelligent behavior". The term "prerational" as applied to describe a "level of intelligence" is not currently used in psychological studies, and for obvious reasons.

First, numerous attempts from experimental psychology to measure levels of "intelligence" by IQ (Binet & Simon 1916) or to analyse its basic components by factor analysis have not fulfilled expectations for a better understanding of these mental capacities and they can hardly contribute to clearly demarcating rational versus prerational behavior. When asked how he would define "Intelligence", Alfred Binet, the promoter of the first attempts to use a standardized method for testing it in children, answered "I simply call intelligence what is measured by my tests".

Second, proponents of the new approach to robotics concede that the scaling of a robot's real-world performance in terms of "behavioral intelligence" is far from clear (Brooks 1991), although several ambitious tests are in progress. More generally, the behaviorally centered approach of "intelligent" productions by autonomous systems promoted by "Animal Robotics" (McFarland & Bösser 1993), and presented by some as a paradigmatic revolution, would, at first sight, probably be considered by psychologists merely as an engineering-minded neobehaviorism. At one time, classical behaviorism in its basic principles also claimed to lead to a radically new understanding of mental processes, but unfortunately it has been unable to fulfill its expectations. Therefore we would expect from psychologists some a priori reservations with regard to this new approach, which will be suspected from the outset of simply rediscovering and reformulating, sometimes in a naive way, old well-established psychological notions.

Be that as it may, the endeavour deserves more attentive consideration from at least two different points of view. It might first be considered rewarding if its primary goal is to develop working control systems for autonomous mobile robots and then to design more efficient machines able to increase human control of the material environment. Second, and much more challenging, however, would be the prospect, sometimes claimed by the proponents of this new field, of helping psychologists to understand how evolutionary and environmental constraints have permitted the emergence of rational behavior from the complexification of simpler prerational organisms. In other words, the main problem would be to trace the links in the chain from "low-level" sensorimotor mechanisms up to "high-level" cognitive activities in order to understand how increasingly complex biological machines have been endowed with thinking brains able to self-generate the logico-mathematical rules that guide their symbolic reasoning.

The later, of course, has always been of central concern for past and contemporary psychology in its various fields of approach. Several aspects will be presented here, all emphasizing the relation, in human behavior, between "low-level" sensorimotor mechanisms and "higher-level" cognitive operations that interact in organizing intelligent adaptation to environmental constraints.

Two main problems will be presented here: the first concerns a modern reappraisal of developmental approaches in the constructivist framework of Piagetian epistemology. The study of the transitions in children between what is termed "practical" intelligence of the early sensorimotor stages and the subsequent "operative" intelligence based on constraining logical rules is at the heart of our understanding of how rational behavior could progressively emerge from more primitive "prerational" mechanisms. The second concerns the coexistence and the cooperation in the mature organism of a cognitive apparatus with a basic sensorimotor machinery and the various ways in which higher cognitive levels may adaptively control or supersede this sensorimotor machinery.

2. FROM EPIGENESIS TO PSYCHOGENESIS. THE EVOLUTIONARY GROUNDING OF INTELLIGENCE
(Guy Cellérier)

Among the theories of the development of intelligence proposed by psychologists (Piaget 1952; Vygotsky 1978; Werner, quoted in Bidell 1988, and Dean 1994), the evolutionary approach of Jean Piaget is probably the most appealing for an evaluation of the new behavior-based approach to robotics (Brooks 1991). Piaget emphasises sensorimotor interactions between the organism and its environment as a necessary but not sufficient condition for building the architecture of intelligence. Similarly, behavioral-based approaches to robotics stress the need to introduce, in the machines, active processes and dynamic interactions with the real world but, unlike Piaget, consider these as both necessary and sufficient for building self-organizing intelligent systems. The latter see no need to resort to the traditional central abstraction of symbolic representations that is part of Piagetian constructivism.

To make this difference clear, we must begin with a reappraisal of Piagetian constructivism (Cellérier 1992) from a functional evolutionary perspective and in the light of newly acquired knowledge concerning genetic processes and neurobiological mechanisms of epigenesis and learning processes. Developmental models in conjunction with an evolutionary process make it possible to study how genetic information and environmental influences interact and complement each other during development. Evolutionary approaches generally bypass the epigenetic process of development and directly map the genotype into the phenotype. Developmental processes as potent adaptive mechanisms have to be taken into consideration together with evolution and learning in creating autonomous living systems. First, however, we should sort out some of the semantic difficulties raised by the concept of prerational intelligence introduced by Animal Robotics.

2.1 *Rational and Prerational Forms of Intelligence*

From a functional perspective, prerationality is quite rational. For Pareto's homo oeconomicus, the ancestor of our present game-theoretical homunculus, any means of action that is adequate to produce its adaptive goal is rational. If we adopt this means-end conception of Kant's practical reason, as practical or pragmatic rationality (all these terms refer to the rationality of a given performance of an action), then all schemata of behavior that are 'adapted' to the realization of their function or goal are rational. Even the predatory amoeboid movements of "prerational" gradient-following unicellular organisms are thus the behavioral realizations of rational sensorimotor schemata, that realize their practical adaptive function of feeding the animal.

This observation requires some conceptual analysis of the intended meaning of animat or robot 'prerational intelligence'. We do not have the space to do this here; instead, we briefly situate this idea within the framework of the functional architecture proposed by psychological constructivism.

We will consider three theoretical dimensions for situating prerationality: first, the obvious distinction between thought and action; second, the distinction between latent and activated forms of thought and action schemata; and third, the dissociation between a sensorimotor and a semio-operator executive machinery.

2.1.1 *Thought and Action*

The distinction between rational and prerational is ultimately based on the 'natural' products of psychogenesis: the notions of thought and action which are central components of our practical, and constantly practiced, common-sense psychology. This naive psychology does not confuse knowledge and knowing with know-how and doing: look (and think) before you leap (or do) are amongst its maxims, at least for European children (and eventually their experimental reflexive European rats). The first implication of this practical evidence is the cognitivist distinction between 'non conscious', 'preconscious', 'unconscious', 'non-articulate', 'implicit', 'procedural', etc. "knowledge" that forms 'know-how', and on the other hand its conscious counterpart: 'representational', 'declarative', 'propositional', 'predicative' knowledge, or in Ryle's (1949) terms, between 'know-how' and 'knowledge that', respectively. The distinction appears to be between conscious objects of knowledge – Brentano's (1874) often quoted so-called "intentional objects", which really mean 'attentional' objects (conscious objects of attention) – and enactive know-how, which, if it is knowledge at all, is not conscious knowledge. Indeed there are also latent forms, not of knowledge but of action, as distinguished from thought in the theory-in-practice of naive psychology.

Thought and Action can thus be perceived to be the initial intuitive objects of the subtle distinctions between the enactive 'know how' (to do) of external action and the conceptual 'knowledge that' of thought. Ryle's 'knowledge that' is more a 'knowing what', the knowledge of what memory 'reflects' in the mind, in answer to the implicit or explicit "w-questions": who, which, why, wherefore, where, when, etc. These are the multiple temporal, spatial, causal, means-end, possession, etc. categories of practical reason that are communicated in the meaning frames of languages and form the unmarked practical cases of grammars. The first basis of the distinction between prerational and rational intelligence that can now be made explicit on this action-thought dimension can logically be formulated as follows: "prerational is to rational, as the 'know-how' of action is to the knowledge of reflection."

2.1.2 *Latent and Activated Schemata*

A distinction must be introduced here between latent and activated forms of both thought and action schemata. This will be the basis of the second theoretical dimension for our analysis of the conception of prerational intelligence. It opposes the latent form of schemata as constitutive memory in the general sense of unstructured memory traces, to their active form when activated and in the process of producing behavior and conduct: the internal activities or conducts of thinking, and the external ones or behaviors, of doing. Since thought and action are the products of activation, activation of distinct semio-operator and sensorimotor schemata, these schemata do not become identical in their 'latent' form as unstructured memory traces.

Thus unstructured "prerational" action schemata are not knowledge or re-mindable as knowledge. They are thus not even latent or prerational knowledge. They are simply the material signifiers of information. The first theoretical consequence on this dimension is thus that the information unstructured in the latent 'memory trace' form of the prerational schemata of practical sensorimotor intelligence is itself not knowledge, and further that it cannot be directly re-called as knowledge, since sensorimotor schemata do not determine thought activity.

The consequence for rational intelligence is that knowledge is either a conscious product when the semio-operator schemata of thought are active, or it is not conscious when they are latent. But the information unstructured in these semio-operator schemata of rational intelligence is not knowledge in this latent form; it is the unstructured material signifier form of the semio-operator schemata, which when activated will re-mind, thus re-produce (and produce) conscious knowledge in the mind. Though this information will produce knowledge later when the schema is activated, it is not knowledge yet. Thus latent semio-operator schemata are no more unconscious knowledge than stopped clocks are unconscious chronometers, so to speak. They will produce chronometry later, when they are activated. Similarly, unconscious knowledge cannot be both unconscious and knowledge at the same time. Whenever it becomes 'unconscious' as memory, it stops being knowledge, and conversely, when it starts being consciously recalled from memory, it becomes knowledge. Schemata cannot be latent and active at the same time: when they start being one they stop being the other. In sum neither inactive prerational schemata, nor even inactive rational ones, become a special form of knowledge.

We must resist, here, the urge to join the epistemological free-for-all quasi scholastic debate about the existence of "Unconscious Knowledge", and instead turn back to our basic evolutionary perspective. As we saw, whether in its active or in its latent mode, a semio-operator schema is not functionally the same as a sensorimotor schema. In what precisely, then, are they different? Their psychofunctional differences will define our third axis for situating prerational and rational intelligence.

2.1.3 *Sensorimotor and Semio-Operator Schemata*

The information elaborated in these two categories of schemata operates two different kinds of execution machines: the sensori-motor one that executes the action schemata producing behavior, and the cortical semio-operator one that executes the representation schemata producing thought. These schemata and their execution machines are realized by different anatomical structures (see Section 5.3), they realize different psychological functions, and they exhibit different psychofunctional characteristics.

The third implicit distinction between prerational and rational results within this descriptive dimension. It distinguishes the two execution machines, setting up a correspondence between the sensorimotor machinery and the automatic, unconscious, practical schemata of prerational organisms, and between the semio-operator module and the 'cognitive' representation and later reasoning schemata of higher organisms. This correspondence now implicitly bases the distinction between prerational and rational on the common-sense psychological contrast between 'reflex', 'automatic' behavior, and 'intentional', consciously willed behavior.

Within this semantic opposition, automatic takes on the double connotation of non-conscious and non-voluntary: non-consciously willed behavior in which our schemas operate us, thus using us as their execution machines. This contrasts with those of our behaviors that are conscious and voluntary: 'knowingly chosen, decided, and executed by responsible acts of free will'. These acts, in which we use our schemata as slaved or servoed execution mechanisms for the execution of our intentions, constitute the practical conception of 'intentional' behavior. This is the conception applied in affective exchanges, in practical morality and in law (and, it seems, in psychological philosophy). This intuitive distinction between willed and non-willed cannot provide us with a strict theoretical dichotomy (between non-willed as prerational, and willed as rational) because the willed category contains both willed intentional, and thus non-automatic behaviors, and automatic (including non-rational), and thus non-intentional, thoughts. Thus we again need to 'analyze out' the undesirable semantic connotations of practical psychology, instead of importing them in 'uncritical' implicit form into psychological theory.

To sort out the reasons for this psychofunctional tangle we return again to the evolutionary perspective to reconstitute how that perspective can interpret the progressive evolutionary rigging and wiring of new types of learning machines into the existing ones.

2.2 From Epigenic to Psychogenic Formation of Schemata. Three Overlaid Learning Execution Machines

Psychology emerged on the planet when some genes evolved behavior, as a new way of 'adapting' (this is shorthand for the sentence: "...multidimensionally boosting the differential reproductivity of...") their carrier organisms. This eventually led some other genes in multicellular organisms to evolve, and store in the very small space of the new neurosynaptic microcircuits, a far greater number of new potential adaptations than exclusively anatomical change could support. Moreover, this new wealth of evolutionary potential is accompanied by an informational economy, because it is no longer obtained primarily by changing the basic anatomical-physiological structure of the organism but only by changing its behavior schemas (and eventually the latter's sensorimotor execution sub-machine).

This new form of adaptation, behavior, is what functionally separated the first animals from the plants. This happens when primitive schemas start using the same organism as their common execution machine. Purely 'instinctual' schemas are the ancestors of our present 'automatic' schemas. This observation sorts out the first part of the psycho-functional tangle. Automatic schemata of rule-governed internal activity already existed, however, in the form of what we called the internal homeostatic 'physiological instincts'. And these 'cellular metabolic' schemata are ultimately instantiated at the genetic level. Thus schema-based behavior in itself is not what distinguishes psychological activity from physiological and even genetic activity. This is because the essential difference is not where the behaviorists are looking, but one level of description higher: the difference exists not within the isolated S-R behaviors themselves, but between them, in their coordinations: in the sequential form of division of labor that characterizes the cooperation of behavioral schemata, in contrast with the parallelistic cooperation of physiological schemata.

To put it briefly, physiological activity schemata are built into their own execution machines, variously called organs, organelles, etc. the lower they are in the vertical division of labor hierarchy. They can thus operate, and also evolve, simultaneously with other organs, and hence also cooperate in parallel mode.

Part of the advantage of behavioral schemata resides in this sharing of a unique common 'all purpose' (the purposes of the "four f's of biology: feeding, fighting, fleeing, and reproduction" to quote Michael Arbib (1972)) execution organ: the sensorimotor system of the organism. But the less advantageous consequence is that all the behavior schema specialists cannot operate simultaneously, thus their basic mode of co-operation is sequential and not parallel.

Thus the fundamental problem for the behavioral division of labor is setting up sequential and periodic agendas (from daily routines to seasonal ones), allocating the temporal order of tasks, and distributing them over time for time sharing of their multiple specialized schemata on the common execution machine. These sequential forms of organization of different cooperating specialists diverge from the parallel synchronized forms of organic activities, and this evolutionary move from synchronizations of simultaneous activities to time sharing of successive ones is what makes behavior the indirect origin of psychology.

By focusing on the temporal structure of behavior we can discern a unity between its motivational affective component and its means-end intelligence one. Both appear as agenda formation mechanisms, but act on different time scales. These scales correspond to the slower inter-task successive schema-team activations of the first, and to the faster successive activations of intra-task cooperating schemata within these teams.

The first behavior schemata are, as we saw, 'wired' into the neural circuits by the same epigenetic processes that form the other organs. The classical 'instincts of conservation and of reproduction' which they implement, realize the first external behavioral servomechanisms of "the four f's" mentioned above: the first three are the previously exclusively internal organic 'homeostatic' regulators, and the fourth is the internal cellular (meiosis, etc.) reproductive servo-machinery of the genes.

For the same reasons as those for which the 'innate' internal epigenetic migrations of cells are not exclusively ballistically pre-informed, the now behavioral innate external migrations of their societal wholes, the organisms into which the cells have recursively regrouped during morphogenesis, are also assisted by post-information. Thus the 'acquisition' component also appears at the same time as the 'innate' one at this new level of the external molar behavior of the whole. Moreover this is practically an immediate evolutionary step, as it is a direct extension into this new "learning" function of the same induction-based instructing mechanisms that already serve to anatomically pre- and post-form the instinctual schemas' circuits themselves. Since all learning ultimately comprises a 'memory tracing', focusing on this component does not allow us to distinguish the distinct types of learning processes and products from which these traces result. The consequence is that the early secondary circuit formation, reentrant neural map correlation, parameter setting etc. and the different subsequent stage-specific psychogenetic processes of schema formation are all called learning, and treated as similar.

The three central functionally distinct categories of learning suffice to characterize our three different execution machines (physiological, behavioral and psychological).

2.2.1 *The Epigenetic Sources of Schemata Formation*

Most evolutionary models of biological organisms bypass the epigenetic process of development and map the genotype directly into the phenotype. Epigenetic mechanisms intervene between genotype and phenotype as a potent adaptive process. This mapping is not abstractly conceived as taking place in a single instant. Instead, it is a temporal process that takes a substantial portion of an individual lifetime to complete, and which is highly sensitive to the environment in which the individual develops. A given genotype is never translated into "the phenotype of the corresponding adult organism", but rather into a class of different phenotypes that form the "reaction norm" of the genotype. Our first mode of learning, epigenetic parameter setting, is characterized by the mechanism that selects from the range of the reaction norm of the phenotype that which best fits the requirements of the ambient environmental constraints.

2.2.1.1 *The first prerational schemata were the parameter setters.*   This mode of learning reduces to finding the proper values among a predetermined set of alternatives for a given parameter. Finding this value is the acquisition part of this learning: its post-information component. Setting the parameter to this value is the instructing or memory-trace formation component. The characteristic object or product found and formed as content in this type of learning, hence what is 'to be unstructured in memory' as the 'memorand' (operand, or argument) of the memorization process, is thus a particular value for a setting.

More generally, as we saw in the case of the reaction norm, the memorand is a vector of such values. Imprinting is the homologue at the behavioral level of this epigenetic parameter setting. Since mother hens come in varied plumages, colors, sizes, etc. the chicks are furnished with a small predestined learning space formed of a subset of values on these descriptive dimensions. Ideally each phenotypically possible hen should occupy a point in this space. Imprinting the coordinates of one point allows the chick to discriminate (and selectively follow) the corresponding hen. The canary song example can be simply assimilated to this process, that is, locating the parent's song in a restricted space of song descriptions. Chomskian language learning has much of this flavor, so it would follow that we acquire our parrot language in the way that canaries learn their song.

In summary, this kind of parameter setting learning must exist at all levels in the Tinbergen language hierarchy of the various learning processes of the language levels, but not all language acquisition reduces at all levels and on all dimensions to this one process. Learning a lexicon, for example can proceed without parameter adjustments.

On the motor side, as we noted, parameters must be constantly set and reset during growth, and more extensively during metamorphosis. This possibility of constant resetting differentiates the general form of parameter adjustment learning from imprinting, in which setting is limited to a 'critical period' and irreversible thereafter.

We shall examine the memorand's formation under the next heading, but some observations about it remain to be made. This type of learning process forms no new memorand objects, no new phenotypic structures and no new action or recognition schemata. It merely selects one among many predetermined options. A generator of new structures exists, but in both cases it is outside the sensorimotor or epigenetic learning execution machine, in the genetic system. Thus while the exclusively instinct-controlled lower organisms certainly exhibit a behavioral intelligence which is literally prerational, because the source of its rationality is not inside it yet, they are not models for ideal animats because they are not autonomous robots in the strongest sense. Because it relies on genetically given information, a single organism cannot adapt to new conditions. The adaptation can take place only in the course of Darwinian evolution. Ironically, we might add, they are perhaps better images of the present state of the art in AI, since there is the same division of labor between the programmer as an outside genetic system, and the learning program as essentially a multidimensional parameter adjuster for the schemata pre-encoded into it. Today as in the famous eighteenth century mechanical chess player, the programmer is still very much present under the table, pulling the now invisible, and more and more indirect, we hope, software strings. The puppet chess player thus offers us the caricature-like theoretical instantiation of the extreme in heteronomous robots, those whose source of autonomy lies entirely outside them. Again, the theoretical point behind this observation is not that such learning is psychofunctionally trivial or theoretically without interest. On the contrary, this perspective makes it appear as the ultimate functional zero-level, intra-schema self-adjustment component, necessarily present at this basic level in all types of post-informed learning execution machines, at all levels of schema formation hierarchies.

Within this perspective the epigenetic formation of circuits, and reentrant connections between them, can now be perceived to be forms of parameter setting: one connexion out of a restricted predetermined set is found and unfolded. In correlative connectionist learning, one within a predetermined finite set of connexion graphs between a finite set of units is post-selected through the adjustment of each of the individual units' parameters, its connexion weights, by an automatic reinforcement program, itself predetermined by its programmer in the genetic system role. This is the fast neuromimetic homologue of epigenetic circuit formation, and can also be seen as a form of parameter adjustment learning. It follows that a major proportion of the learning studied by behaviorism consists in the setting of a parameter by its instructuring as a conditioned link.

Studying the conditions of the formation of the links alone, one cannot distinguish whether they support intra-schema parameter values or inter-schema structural linkages. Since every schema needs constant adjustment, behavior forms a higher frequency of intra-schema value (re)bindings than of novel inter-schema linkages.

Thus, what empiricists observed was not the filling of the tabula rasa, a completely blank slate, but the completion of a totally innately prefilled multiple choice slate, with blanks left only for picking one of a set of pre-determined answers. This is perhaps the most predestined possible form of predestined learning on the learning continuum.

2.2.1.2 *Early stages of the development of sensorimotor intelligence.*
Finally it follows from this perspective that in the early stages of the development of sensorimotor intelligence most of the infant's activity must be driving the initialization of the parameter bindings of what Piaget calls 'reflex schemata' to form their functional secondary circuits.

There is a progressive 'centrifugal' flow of exercise and control from the coordination of 'globalistic' movements of both arms which successively opens the embedded subspaces of progressively finer movements of the more distal articular extremities. This flow implements recursive parameter adjustment of the innate repertory of sensorimotor primitives which subsequently serve to compose all the acquired schemata that constitute sensorimotor intelligence.

The later formation of the polysensory object as a permanent object, and the reciprocal coordination of vision with prehension, would then implement the functional neural connection of groups of neurons simultaneously activated in different sensory and motor maps during the later stages of sensorimotor parameter tuning. This predetermined functional architecture, in which two somatotopic neural maps are reciprocally connected anatomically (called 'reentrant mapping' by Edelman 1987) predestines the learning, i.e. the detection and structuring, of spatio-temporally correlated neural activities occurring on different predetermined maps. The co-occurrence happens when different qualitative dimensions (smell, sound, shape, color, etc.) are simultaneously activated on the maps of different sensory modalities, or sub-modalities, and must eventually be attributed as distinct aspect properties to a unique common entity invariant over modalities, the polysensory object, and not to multiple independent sensory events. The same type of correlation, this time between oculomotor and manumotor maps, must then be set up during the infant's later directed prehension exercises, to coordinate the separate aiming schemata of the eye and hand, on these polysensory and now polymotoric objects.

This 'staging strategy' opens a new functional subspace only when its predecessors' parameters have been set. It is a powerful general 'psychogenetic anticombinatorial', a problem subdivision, and thus a division of labor heuristic.

2.2.2 *The Behavioral Sources of Schemata Differentiation*

The succession of psychogenetic stages apparently 'recapitulates' the phylogenetic one, simply because the opening of these functionally embedded psychogenetic subspaces is itself staged, but for the anticombinatorial reasons mentioned earlier. Here, the psychofunctional constructivist argument includes and extends the classical structural one about logical prerequisites. It does not only point out that before you can build brick walls you must first produce the bricks. It complements this by adding that if primitives (including material bricks but also more abstract building blocks) are to be functionally adjusted internally while they are externally combined into functional configurations, the compositions of their two simultaneous search spaces would become multiplicative.

We limit the reformulation of the psychogenetic stage sequence in this perspective to focus on the two central functional transitions that span it. The first functional transition starts with what Piaget calls the formation of the first habitual schemata. It is characterized by a 'reconstruction and extension' of the products of the preceding stage, the now adjusted reflex schemata. We use this initial instance to typify the formation process of the whole first transition.

To do this we must first attenuate the hard computational connotation of 'primitives', as building-block like combinable components. At all stages and thus at all levels of schema genesis the bricks of ethopsychological schemata are recursive (sub)schemata.

The highest present level of schemata is the subschema of no new ones, yet. It follows that all lower subschemata are stage- and level-relative primitives, and the lowest, zero level psychogenetic primitives are thus also schemata themselves, but schemata which, like the classical reflex loop, have no psychogenerated subschemata, only inherited ones.

As we observed earlier, however, even reflex schemas are already servomechanisms, that is autonomous division of labor specialists 'slaved' by specific signals or 'activity triggers'. Their function is not selection itself, but the autonomous execution of the task selected within their predetermined domain. These 'soft' signaletic primitives of ethopsychological schema compositions are not passive combinable building blocks, nor are they ballistic-program computational primitives that can be composed like mathematical functions. As informational machines, schemata are not only objects of this functional category of mathematical computational machines of automata theory. If effective they 'are' by thesis formalizable as Turing machines. They also form an extension of the functional category of 'automatics', the practical automata of servomechanisms, automatons, robotics, or perhaps 'bionics'. At this new level the psychogenetic formation of schemata is not 'ballistically engineered' as the composition of mathematical functions or program instructions, nor as that of material mechanical, electronic, chemical etc. building blocks. It is instead in direct functional homology with the evolutionary and morphogenetic formation of multicellular organisms by the migrations and recursive regroupings of unicellular specialized microorganisms. All such evolutionary products as biological, cultural, and psychological entities are, ultimately, evolutionarily stabilized 'societal organisms'. Thus, just as organisms are societies of specialized single cells, schemata are formed when the agenda division of labor links the activities of specialized subschemata in teams cooperating on the same task, becoming unstructured or conditioned, SR-linked, associated, etc. The new, now compulsory, schema 'association' is thus a societal entity formed of autonomous microservomechanisms slaved to the realization of the new function of the whole.

2.2.2.1 *Learning by refining subdivision of labor.* The process of forming these associations is logically analogous to the formation of the succession of temporary ant teams that briefly form, transport a prey a small way and dissolve, to be replaced in a short time by a new team. The difference is that the ants' algorithms are fixed, while the schemata are flexible.

Common sense intelligence is the result of the progressive psychogenesis of multiple pools of functionally co-adapted specialist autopilot schemata, whose constantly improvised cooperations pilot us through the routine 'everyday life' tasks, that are in effect our 'everyday multiple sublives' with their corresponding specialized professional, practical social etc. task subuniverses. This succession of temporary improvised ant-like schema teams, each of which moves us a small way along the day, forms what is, in this conception, more our intertask agenda of time sharing of our multiple common sense, behavioral, and other intelligences, than the partless whole which our psychological common sense has us perceive.

Within this psychofunctional conception, when a task repeats sufficiently for reimprovizations to become more costly (in effect and effort) than noting and repeating, differential access to members of more efficient teams is initiated. The elect team becomes the 'memorand' component of learning. The memorization results from the instructuring of its interspecialist access links, to become a new whole: a new 'societal being' superspecialist.

This top level access completes the team's formation and institutionalization in the pool as a new psychological unit: a psychogenerated 'chunk', 'macro' or 'schema' which can now interact with other preexisting schemata. Legal systems form new composite societal 'legal subjects' in much the same manner, by the founding of the only artificial immortals in society: the joint-stock companies who survive their more ephemeral human shareholders.

The formation of the memory trace by rehearsal and repetition is thus not the automatic tracing of a repeatedly beaten path proposed by associationist theories. It is on the contrary an instrumental process for the differential conservation of schemata, an internal memory-schema actively triggered under certain conditions by the acquisition and filing strategies of intelligence.

Before we move on to the formation process that produces these first composite habitual schemata, we need to add its vertical dimension to the traditional, implicitly horizontal specialization evoked by division of labor. The traditional sense focuses on the horizontal 'blue collar' specialization of the cooperating specialists: mason, carpenter, plumber, etc. for instance who directly execute their intertwined subtasks in building a house. The vertical division of labor is that of the white collar administrators, the agenda makers and dispatchers who only push paper, not timbers. The focus on the horizontal subdivision masks the existence of the vertical one even in the original, administrator-free example. Each of the participants in the building of a house is invisibly subordinated to this common end, the realization of the plan. Here as in our previous gene teams, the control of subtasks is distributed among the specialists. Thus part of their activity is vertical administration, and together these parts form the invisible non-centralized agent and time distributed administrator, the virtual supervisor. Coordination by subordination to a common goal is itself divisible, but must recurse downward in that each agent, when necessary according to the task's operational decomposition, must subordinate his activity to realize another agent's subtask. This 'submeans' subordination relation is of course mutual: both agents may repeatedly call each other reciprocally for their own current task in the subtask stack.

From the functional perspective we can distinguish two extreme opposite ways of forming new divisions of labor. One consists in starting with functionally specialized parts and integrating them into a whole with a new function, baptized 'emergent' but only at the moment of its completion. The other consists in starting with an existing structure, using it as a 'prototype model' to fill a new role in a 'germinally' functional or 'undifferentiated' manner, then propagating this top level protofunction downward to differentiate it in progressively more productive specialized parts. Again, the part integrative method is preinformed by design and engineering, while the differentiation process is postinformed by the external information 'back feed' of evolutionary post selection, and it is based on repeated downward differentiating by the vertical submeans that we have introduced in the concept of division of labor.

Each successive form must improve the function of the whole on one or more evaluative dimensions, thus differentiating one or more subfunctions.

We postponed functional analysis of the generative process of primitive innate schemata until this point because its basic genetic (re)productivity-governed variation and selection cycle is functionally homologous to the first form of psychogenetic productivity trial-and-error learning which starts with the formation of habitual schemata.

2.2.2.2 *The formation of habitual schemata.*   From the perspective we propose here, the "reconstruction and generalization of reflex schemata by the incorporation of new objects" which forms habitual schemata is a process of 'acquired differentiations' that uses the 'innate' reflex schema as the organizer of a new schema for a new function. This 'innate' already functional framework serves in a second phase to organize and hence assist a later acquisition process. It is thus another form of realization on the continuum from predestination to learning (Minsky 1986). The same process recursively reapplies to (and within) new habitual schemata: once formed they can in their turn be tried as protoschema candidates to new functions by "assimilative generalization of new situations to similar familiar ones". This assimilation repeats the protoschema's new parts (its subschemata) to recursively differentiate it into forms of final division of labor as elaborate and 'behaviorally intelligent' as those of the preceding instinctual schemata.

The function of extending differentiation culminates with what Piaget (1983) describes as the psychogenetic limit between habit and intelligence, when habitual schemata such as repeatedly pulling on a string that has the effect of dangling distant hanging toys are triggered by the discovery of a newly inserted toy. "...this conduct is close to its articulation in a means (reconstituted after the fact), and a goal (set after the fact)." Indeed we may notice that it is not, as it was initially, the sight of its nearby string antecedent that triggers the schema, but the sight of its consequent event's site and (toy) scenery. The present site of a future event becomes the trigger, and thus becomes a signifier, a present material index pointing to a future goal, an external material representation.

The next step leads to representation of the goal before the search for the means. This is the beginning of the semiotic or representation function, and coincidentally both that of true sensorimotor intelligence and of intentionality. "From then on, the goal is set before the means, since the subject has the intention of seizing the objective before having that of removing the obstacle," writes Piaget commenting an observation of this stage of the prehension schema.

This early entry of representation as the central factor in the development of sensorimotor intelligence excludes all but its initial stages as instances of 'prerational' or purely 'behavioral intelligence', since all the later stages of this 'practical intelligence' develop the representation of goal situations and of means of action to attain them. This component transforms their schemata into 'reflexive intelligence'.

Finally we may observe that the articulation point between habit and intelligence is simultaneously that of our second functional transition. It leads from the mode of learning which we have just examined, a mode in which the formation of the memorand results from the automatic behavioral differentiation of proto-wholes, to a mode where its formation results from intentional assembly of wholes from primitive parts.

Before we move on to this next mode, we introduce functional analysis of the common, gradient-following generative process of both innate and habitual schema differentiation.

2.2.2.3 *Generator schemata following productivity gradient.* The reelaboration at each evolutionary level of the simple and economical gradient-following device is an Ariadne's thread that leads from the phylogenetic level through the epigenetic to the top psychogenetic levels. Some of its aspects are clear candidates for prerationality. Accordingly, gradient following is not the optimal strategy. Any random or systematic trial step leading the system from its present position to one in which an increase in concentration is detected, is 'positively selected' by being substituted as the new departure point of the next generation of trial steps. (Here, non-selection is negative selection or reinforcement.) This substitution of the last arrival point as the next departure point thus simply cyclically reenters the schema into itself.

The first prerational aspect of this generator schema is that any step that ascends the gradient, or equivalently 'majors', maximizes, optimizes, etc., is selected. This schema does not guarantee the shortest way up to the top of the hill, the straight 'line of steepest ascent' (or conversely for phobias) from the present point to the top of the hill. This schema's immediate selection of the 'first better' (major) generated trial is thus not equivalent to the rational ideal , the "best first" method of differed selection of the best (maximus) of the set of all possible trials of a systematic and exhaustive generator, which would always lead up to the top along the unique straight, and shortest line of steepest ascent. The random trial generators need far less information to specify, and are thus most frequently formed in majorative schemata.

With such generators the schema will eventually lead to the top, but along one of the set of all combinatorially possible longer 'monotonically increasing' random walks. This set can be huge and it includes only the shortest walk as a rare special case, which is precisely that of means-end rationality sought by optimizing, minimax, etc. strategies. In a sense, then, the most frequent form of the schema increasingly guarantees with increasing search space size, the very opposite of the rational optimal one: on the average, the shortest path ideal is not achieved, and less and less frequently so. This is a first indication of prerationality. Multirealizability of means to reach the same value is compounded at the next realization level because the same value may be achieved by several routes, and further again because in multidimensional value spaces, different structures with different values on different dimensions may achieve the same 'weighted average grade' value.

The function of the generation of a subgoal is to detour around an evaluative obstacle in the form of a downhill move away from the main goal. This final observation about the irreversibility of the differentiation schema again leads directly to our second functional transition, toward representational intelligence, whose central character is precisely what Piaget describes as the rational reversibility of its operations. In contrast, non-reversibility and other aspects of the non-representation driven psychogenic process of the habitual schemata of the first stages of sensorimotor intelligence, would now indicate its prerationality in the Piagetian framework. Finally, we can examine this theoretical possibility as we previously did for the instinctual schemata.

As we already observed, purely instinct-operated animals are not yet autonomous robots because their schema generator is outside their execution and learning machine, in the genetic system that plays the role of the software puppeteer pulling their signaletic strings.

The phylogenetic extension of parameter learning into differentiation learning now transfers the lumbering robot's puppeteer, the autonomous schema generation process, inside it, into its behavioral learning execution machine. The selective advantage that allows this transfer is again not uniquely determined. The delegation of this function offers an escape from the genetic 'multiplicative' combinatorial explosion that results from 'adding' more and more new instincts into already numerous highly evolved and complex ones, and will inevitably reach a point where it will functionally paralyse any finite generator. A second advantage of this delegation is more obvious since the formation time of new habitual schemata is counted in the 'fast time' units of the organism's life span, while that of new instincts happens on the generation-to-generation 'slow time' clock beat of its genes' mutation and recombination. The obvious combinatorial advantage is that the trials per generation and the productivity of the schema generator climb very steeply.

These first animals that learn exclusively by automatic trial and error are the only instances of pure behavioral intelligence. Their schema differentiation process is "prerational" in relation to the assembly-type processes of later "rational" planning and design. Have we then found the model precursor for prerational 'behavioral intelligence' animats?

There are two main reasons why this is not the case. First, automatic differentiation of schemata is not exclusive to behavioral intelligence. The schemata of representational intelligence undergo the same automatic differentiation, from the same prerational gradient-following generative process. Poincare "incubation" is a manifestation of this permanent fundamental psychogenetic process, which underlies the earlier formation of Piaget's "preoperational" notions and schemata of representational intelligence. Thus prerationality cannot be simply identified with behavioral intelligence. It is possible, though, that prerational processes can be a useful approach to the development of rational intelligence.

The second reason touches on functional architecture. Lower animals have two co-operative schema generation machines, working on two different time scales. If we merely 'lift' or 'skim' the early behavioral learning and execution machine into an animat, we cut it off from its genetic source of schema formation. The reason for this is that only the genes combine new primitives. The behavioral schema differentiation system does not. In consequence it may exploit a gene-produced protoschema repertoire, but it cannot renew it, and thus it can neither renovate nor innovate behavior. Such initially autonomous systems are condemned to evolutionary stagnation by the very incompletion of their 'lifted' functional architecture, after an initial productive phase. Their indirect autonomy must then be replenished by an outside source, the genetic system, or the programmer now acting in this role.

Programmers, however, have developed a way of playing the same role of renewing the protoschema repertoire, but from above. What allows them to do this is that our own system of schemata differentiation has evolved a second external source, in our representational system. Programmers use this planning system to play their genetic system role, artificially imitating the genetic algorithms of nature. Its evolution also constitutes our last functional transition with the introduction of our third learning mode.

2.2.3 *The Representational Source of Schemata Combination*

Representation plays a role in the later stages of sensorimotor intelligence, on the automaticism and non-reversibility of the initial stages of the psychogenic process, including an automaticist differentiation during the 'incubation phase' of the corresponding semio-operator schemata. This points to the next step toward reflexive intelligence: the functional transition, together with its corresponding psychogenetic stages of implementation toward value-governed self-evolution. The functional continuity between epigenesis and psychogenesis extends to these higher evolutionary levels. From this viewpoint a functional transition can be seen as the formation in humans and some other animals of a new kind of ethological entity, a species-specific 'psychogenetic instinct'. The contra-intuitive particularity of this novel instinct in humans is that, contrary to the common view of genes as blueprints determining behavior, this instinct predetermines the development of final forms of cognitive systems whose central function is the planning and self-determination of behavior. "Genetic" is used in the sense of Genesis, or origin within an individual's lifetime. In short, according to Piaget, in this etho-psychological conception of psychogenesis we are instinctively predestined to the learning of self-determination. We are of course also synergistically sociogenetically reinforced to become effective moral and legal adults 'responsible for our own conduct'.

This concept should make it apparent that the phylogenetic ethological and psychosocial notions and problems of self determination are not reducible to physical determinism, or to quantum or other indetermination, in the brain. The underlying biochemistry of the brain of the master, human or ant, is as little physically determined, or indeterministic for that matter, as that of his slave; and this continues to be true intrasubjectively for each of them separately, whether the 'master' or the 'slave of his passions.' Even non-deterministic systems can be hierarchically subordinate. Here again, as in the case of the behaviorist S-R units, there is a search for higher level properties inside lower level constituents. In many senses, this 'reflexive' form of intelligence is a late development on both evolutionary scales. It demonstrates a final Piagetian (1952) "formal operatory" stage in children, at what was socially considered to be the age of moral and legal consent, and thus also the 'age of reason'. We abstract some of the functional characteristics of this stage to make explicit, and thus situate, this theoretical extreme of the continuum from prerational to rational.

2.2.3.1 *Assembly and coordination of means by subgoaling to common ends.*
We now pick up the stage sequence where we left it, at the psychogenetic

articulation point between behavioral and representational intelligence, just after the child has differentiated the string-toy configuration puppeteering schema into a means-end couple "reconstituted after the fact". His next step, to seize a toy when the child must first set aside an obstacle which masks it partly or totally, is the practical paradigm of the next stage. The schema of seizing the objective "assigns a goal to the action". This in turn induces the status of "being used as means" to the schemata of hitting, pushing, etc. the obstacle. In this manner different schemata now become "susceptible of being coordinated with one another".

These schemata were formed before the operation of representation, for different functions, on different objects and situations in space, and at different times during development. They must now be evoked from their latent memory form to interact in a coordinated division of labor, for common function in the same work place. This now 'goal driven' mobilization by internal signals, of schemata that previously were mobilized or 'stimulus driven' only by external situations, is the new co-ordinative control function of representation. Phenomenology and subjective qualia of consciousness become immaterial from this control perspective. Whatever the nature of the qualia of conscious contents, they will realize the control function of representation. But correlatively this new kind of control has revolutionary effects on the organization of schema combination. These transformations of control flow (and thus of the corresponding 'flow diagrams') cannot be dismissed as mentalistic epiphenomena: they are physical events in the brain. Ironically perhaps, one of the tricks evolved for representation consists precisely in tricking an internal schema into spontaneous recognitive self mobilization by presenting it with an internal lure. Such signifiers, called mental images, hallucinations, imaginations, etc., consist of higher level descriptors in the recognitive schemata of the external objects of selected sensorimotor schemata, which are thus 'internalized'.

The 'evocation' devices of representation now allow schemata from different origins to enter as components for assembly into improvised wholes, which may habitually associate into new 'societal' division of labor teams. As we saw previously the customary divisions of labor become codified army divisions and units when they finally are accessed as unitary schemata.

2.2.3.2 *Functional subgoaling and spatio-structural detouring.* The new ascending process of designed assembly is revolutionary in that it is the functional opposite of the previous descending differentiation of wholes into subschema parts. This assembly process is directed by the progressive formation of a new schema-composition system. This societal system acts upon schemata as its objects and thus forms a new solution generation method formed by the solution generator system itself. The mechanism of this schema was later captured by Newell and Simon (1972), characterized as "subgoaling", and programmed to form the functional recursive control core of their General Problem Solver.

To pull the two approaches together here, we use Piaget's hen to analyze subgoaling. This hen's practical impasse was often used by Piaget to concretize the (algebraic) structural aspects of the conduct of detouring around an obstacle. The functional structure of the schema of this behavior is the homologue in its comparative behavioral anatomy with that of the child's subgoaling manipulation schema on a masking obstacle. In its headless chicken scenario, the hen is let loose, and makes a beeline toward grain placed some distance away. However the experimenter deftly lowers a fence perpendicular to its projected beeline. This instantly realizes an abstract gradient-follower theory trap, because if we choose the distance between the hen and the grain as an evaluation function of the improvement produced by its last step, any step toward the edge of the obstacle to detour around it increases this distance and is thus not selected by the 'first better' gradient ascender schema. The hen should in theory oscillate around the center point of the obstacle, and it is observed to do that, to the satisfaction of the animal psychologist, at least. For Piaget this absence of the "conduct of detour" in the hen is a perfect illustration of the lack of some theoretical algebraic properties in a proto-structure of practical space constituted by the (re)grouping of the animals' schemata of locomotion. From the psychofunctional perspective, we would however observe that the hen's detour problem lies at a deeper functional level than the properties of its "general coordination of action" laws in space. The reason is that these properties are defined on the composition of actions, and the detour problem is itself a problem of composition, not of its structure defining properties.

The fact that composition produces a new whole of the same nature as its parts was often called 'closure'. This qualitative re-injection of the products of the composition of its components into the original component pool produces a property of indefinite recombinability, which constitutes such element pools in mathematical structures. Thus adding or multiplying positive integers always produces positive integers as results, and so generates qualitative closure; subtracting or dividing do not, and they are correspondingly not structure-producing schemata. The indefinite recombinability of closure is a criterion for psychogenetic improvisation of regroupings of sensorimotor and later semio-operator schemata, selected on the basis of this indefinite productivity to form the logico-mathematical protostructures which Piaget defines as "action or operation groupings".

In this context, closure gives an indication of where to look for answers to the central epigenetic question: "how do the cells know with which cells to regroup?" at the psychogenetic level of the schemata. How does representation accomplish this fusion? This question opens directly onto the functional core of GPS, Newell and Simon's own artificial theoretical hen: "A Program That Simulates Human Thought". The difficulty with this is that evolutionarily these two agencies are reciprocally subordinated. They later evolve their own autonomy of self-maintenance and self-amelioration, but initially the selective advantage offered by faster, less materially costly and risk-laden "representational trial and error" over the external trial and error of 'behavioral intelligence' is decisive for its evolution. The two agencies are linked by the biologically given capability of the organism to develop behavioral intelligence in interaction with the world.

2.2.3.3 *Sensorimotor and representational intelligence as reciprocally subordinated.*   Organisms that can plan their action, then execute it by subordinating their sensorimotor agency to its plans as sequences of consumatory situations, have an advantage. But if this "subordination control relation" is suppressed they cannot execute their plans and this advantage vanishes. Conversely if the planning agency cannot be subordinated to the behavioral consumatory situations as goals for its planning, the advantage also disappears. Hence the initial evolutionary formation of representational intelligence, and the maintenance of its function, both depend on the existence of this reciprocal functional subordination. These functions are presently and evolutionarily inseparable: trial and error planning gives selective advantage over behavioral trial and error. But directly evolving higher level intelligence from behavioral intelligence is not possible without the existence of some functional architectures. Both functions must cooperate in a division of labor to realize the subfunctions of representational planning and practical execution of plans.

We may capture the essence of this relation of 'planning', 'simulation', etc., by observing that, within it, sensorimotor schemata retrieve, recombine, and store external objects, while semio-operator schemata retrieve, recombine and (re)file signifiers for these perceived objects and actions.

In conclusion, the knowledge-how component of prerational intelligence cannot simply be identified with Piaget's (1983 for instance: the psychology of intelligence) practical sensorimotor intelligence, and correspondingly the knowledge-that component of rational intelligence cannot be identified with rational thought or with Piaget's representational formal operatory intelligence.

3. CONSCIOUS VERSUS UNCONSCIOUS OPERATIONS
(Bruce Bridgeman)

The evolutionary approach developed in the previous section can be applied even to the highest levels of human functioning. Among the greatest challenges to psychology is the explication of consciousness and its functions. Given that consciousness is a rich and complex phenomenon, we cannot assume that it evolved by accident. Instead, we must investigate the functions that preserved and enhanced it over the course of human evolution.

3.1   *The Function of Consciousness*

In this view consciousness must have a function; it must somehow improve human biological fitness. Even if it is considered to be an epiphenomenon accompanying other mental processes, the functions of those processes must be explained. Conscious processes can be either rational or prerational; their essence, in the present functional context, is that they control coordinated behavior that has access to experiential memory and in turn informs new memory. Consciousness is one of the central problems that led to the founding of modern psychology in the 19th century. The original method for studying consciousness was introspection, or looking carefully into the contents of one's own experience to discover its organization. The technique ultimately failed because the data of introspection were internal experiences rather than objectifiable events. Since the data could not be verified, the method was abandoned.

By rejecting consciousness as a legitimate object of study, psychology made great advances in the study of perceptual, motor and control processes but lost sight of the issue of the role of consciousness in mental life. Recently, the issue has re-emerged with more objective methods, such as probes of memory by priming and reaction time, and with the recognition that many mental functions are unconscious.

The problem on which the introspective method foundered, however, is still with us. Consciousness is unlike other subjects of scientific study in that it is a private experience. Its existence cannot be proved by objective methods. In contrast a function such as perception, for example, can be objectified by having subjects perform tasks that require perception for their completion. Perception can then be defined in terms of the information-processing tasks that the perceiver performs. No such definition exists for consciousness because it is the experience of an internal state or process, rather than an externally visible activity.

Nevertheless, it has become possible to develop functional definitions of consciousness indirectly, by identifying abilities that require consciousness. Baars (1988) has applied an old method in psychology, the contrastive method, to this problem. Human abilities possible with conscious awareness are contrasted with abilities that are possible without awareness; ideally, the differences should define the function of consciousness. For example, recalling specific past experiences verbally requires consciousness; using past motor learning does not. Throughout this century the dominant meta-theory in brain sciences has been the stimulus-response link, the connection between environment and behavior. The challenge was to explain what happened between a stimulus and a response, and not much else. Modern cognitive psychology and neuroscience retain a similar meta-theory except that internal processes are interposed between stimulus and response. But the diagrams of the cognitive psychologists always have a stimulus at one end and a response at the other. The mathematical modelling approach of the "new connectionism" is the same at a more microscopic level of neural modelling, with hidden units allowed.

This kind of theory is incomplete in that the motivation for activating a particular block-diagram mode of information flow is not specified. For example, we may understand how visual information from an object is filtered, processed, and compared with remembered information about objects, but the theory does not address why the subject looked at the object in the first place. The act of looking was motivated by a plan, an internally held image of an intended achievement (Miller, Galanter & Pribram 1960). The image is an internal representation, not necessarily a graphic visualizable entity. The plan is defined more broadly here than in its everyday sense: it can control a sequence of actions to achieve a goal.

3.2   The Function of Plans

Plans motivate behavior. Their function is to organize the stimulus-response links that control more immediate neural information flow. Thus the block diagram of cognitive psychology usually describes a method for executing a low-level plan, such as a plan to examine an object. The plans become the prime movers, the keys to understanding how behavior is controlled (Shallice 1978). Current models of pattern recognition, recall, attention, etc. become the means to execute a plan.

If the centrality of plans in human mental life is a reality, then it is important to understand more about how and why they are established and executed, and why the control of behavior by plans rather than direct control by the environment should have evolved. The power of plans is that they allow an organism to escape from the contingencies of its immediate surroundings, to be controlled instead by its own needs in the longer term. A simple animal responds to its environment and to its internal states in a machine-like way and has no need for plans. The fly does not make a decision to feed: when food is at hand in the environment and internal receptors are in the right state, feeding takes place automatically. Humans in contrast, must register several plans simultaneously, executing one while holding others in abeyance. Humans typically have hundreds of plans, from small, immediate ones such as eating the next bite of dinner to large ones such as earning a college degree. Large-scale ones of motivation include a pathway from motivating influences such as instinctive needs or selective attention, through plans, to action (Heckhausen 1991). The plan becomes the path from motivation to action.

Organization of behavior by plans requires neurological machinery to support the planning function and its various ancillary needs. There must be neurological devices to (1) make plans, (2) store them, (3) execute them, and (4) monitor them (Bridgeman 1986; 1988). In order to control behavior, the currently active plan must have access to memory and attention. It must link these functions with perception in order to guide action according to a combination of the internally held plan and the external realities of the perceptual world. Here a natural explanation for the phenomenon of consciousness emerges: It appears in the plan currently being executed. Consciousness of events, actions and ideas is nothing more or less than a result of the operation of a mechanism organizing present sensory inputs and episodes from the past into working memory, and it lays down new episodic memories. The reportability of experiences, thoughts and actions, then, is based on an intimate interaction between behavioral control and memory. No new neurological machinery is required, because consciousness emerges as a result of normal perceptual, memory and control processes in the brain.

Though consciousness in this conception emerges from the planning process, not everything that is experienced consciously is related to plans. Perceptions, to the degree that they enter the episodic memory, can affect behavior and engage the mechanisms that generate consciousness. Thus consciousness becomes not a separate neural module, but a necessary result of the operation of the planning process. It is an active process, not a passive state, the necessary result of the planning process taking control of behavior and gaining access to memory and sensory input.

In this context it is meaningless to look for a box labelled "consciousness" in a brain model, or to try to localize it in the brain's anatomy. The operations that make us conscious occur in the context of controlling behavior, and consciousness has no separate existence of its own. Because it is an effect, not a cause, there is no sense in looking for it as a separate entity.

Planning seems necessary to evoke consciousness, but it is not sufficient. Some very routine plans, even complex ones such as driving home along an accustomed route, seem to be executed without awareness, or at least without a subsequent episodic memory of the events. The episodic memory seems so confounded with earlier experiences of the same activity that it fails to be recorded as a separate experience. The routine activity fails to pass the memory test of awareness: we ask whether someone was aware of an event or action by asking whether they can describe it from memory. In defining tasks that require "deliberate attentional resources", Norman and Shallice (1980) indicated that they (a) involve components of trouble shooting, (b) are ill-learned or contain novel sequences of actions, (c) are judged to be dangerous or technically difficult, or (d) require overcoming a strong habitual response or resisting temptation. Executing a very routine plan that does not meet these criteria leaves the planning mechanism free to engage in other activities.

Almost all neural information processing is unconscious; what we perceive consciously is only the tip of a neuronal iceberg (Bridgeman 1992a). Everywhere the neurophysiologist looks, unconscious processing dominates brain function. The receptive fields, anatomical arrangements and biochemical processes of neurophysiology remain hidden not by repression, but by structural limitations on the planning mechanism itself. We become aware only of the plan currently being executed and of the perceptual and motor events surrounding it.

3.3   Parallel vs. Serial Processing

From a machine intelligence perspective the brain is a massively parallel machine, and a great deal of neuroanatomy supports this idea. All nerve tracts in the brain depend not on one but on thousands to millions of parallel fibers, each with a shared origin and destination. This organization implies that information must be coded in a distributed fashion: combinations of activities of thousands to millions of neurons are needed to code, process and transmit messages. Serial processes are also important in brain function, however. The plan-executing module introduced above has the job of taking a parallel organization of brain activity and converting it into a sequence of serial behaviors or subplans. The mechanism is identifiable in all primates, perhaps in all mammals. The classic studies of Koehler (1925) showed that chimpanzees could plan a behavioral sequence with several steps, such as piling up several boxes in order to stand on them and reach a banana suspended high overhead. Even dogs can solve detour problems that require them to move away from a reward in order to reach it eventually. In the course of human evolution this plan-executing capability began to serve communication as well as action (Bridgeman 1992b). A single idea is recoded into a series of words, and those in turn into speech sounds, in the same way that any other plan (idea) is elaborated. In both instances the execution of actions on the wing is accomplished at lower levels. Speech recognition is handled by another existing module – one that normally monitors the progress of plans – taking a sequence of events and packing it into an idea. Again, an immediate working memory is centrally involved in this process. Language could evolve quickly, on an evolutionary time scale, because it was made mostly out of old parts. The sequencing and comprehension mechanisms were already developed for the planning and monitoring of actions, respectively. Only the articulated grammatical ordering system (Bickerton 1983, 1984) had to be added.

Do planning and language share the same mechanism, or did a new planning mechanism evolve to specialize in language? The open and flexible property of the planning mechanism allows for language to add to its functions with little or no change. The mechanism already had the capacity to handle many plans simultaneously, to organize each one, to assign them priorities, and to handle plans for different kinds of actions. These properties imply that there is a single planning system that can handle both linguistic and action plans.

Once the relationship between language and planning is articulated, a number of seemingly unrelated pieces fall into place. An otherwise puzzling aspect of speech, for example, becomes a consequence: the peculiar relationship between consciousness and language. We are acutely aware of what we say and what is being said to us. Many psychologists from Wundt onward have pointed out the privileged position of language in human awareness.

This view is not without precedent. Indeed, even the word for consciousness in English, Russian and the romance languages translates as co-knowledge, those aspects of mental life that are communicated or are potentially communicable. Russian psychology has emphasized the enhancement of conscious function that comes from social cooperation and communication (Luria 1981).

An action plan becomes conscious only in the process of its execution. What is unique about the linguistic action plan that generates speech is that, while its operations remain unconscious, the linguistic sequence can be fed back into the neurological system that normally monitors external events connected with executing a plan. This capability creates a parallel-serial-parallel route from plan-executing mechanisms, and back into a perceptual module that accepts sequences, all without overt behavior.

If internal speech (Vygotsky 1962) is fed back to a language-understanding process without leaving the brain, ideas from one plan can be elaborated and recombined by the sensory mechanism that normally holds the perceptual results of the plan-monitoring process. Thinking can develop through many iterative stages, multiplying its power with each pass. The relationship between planning and consciousness, however, remains the same for linguistic acts as for other planned sequences, those that act on the environment in an instrumental way. Thus the study of language can also illuminate more general human strategies for organization of serial behavior.

4. COGNITION, LEARNING, AND MEMORY
(Boris M. Velichkovsky)

Contemporary research on cognition, learning and memory supplies converging evidence supporting and elaborating the distinction between pre-rational and rational processing in humans. During the first phase of the cognitive revolution the accent was always on the investigation of intentional learning of verbal or at least easy-to-verbalize symbolic information. This led to models of human information processing centered around the notion of "short-term memory", variously named "immediate", "working" and "primary memory" (see, e.g., Baddeley 1986). According to George Miller's classical study on "The magical number seven plus or minus two", the capacities of this type of memory are limited to 5 to 9 "chunks" of material, be they disconnected letters, words or short sentences, i.e. several times more letters than in the first case (Miller 1956).

All these data correspond to well-known limitations of consciousness in its access to perceptual or remembered events (James 1890). With some corrections, which mostly concern additional restrictions on the size of the "magical number", the study of the role of working memory in performing such tasks as mental arithmetic or linguistic processing remains one of the flourishing areas of psychological research. The results consistently demonstrate that voluntary learning and memory are under tight control of rational strategies and conscious knowledge of the subject about her/his cognitive processes, i.e. "metacognition" (Shimamura 1994). Simultaneously it is clear now that this is only a special case of verbally-mediated and sequential processing. Another, not unessential part of human intellectual resources seems to depend on more parallel and intuitive types of processing which also characterize earlier stages of evolutionary regulation of cognition and behavior.

Indeed, with the rise of interest in unconscious processes and phylogenetic roots of animal and human cognition in the last two decades many other important paradigms of study and theoretical distinctions were discovered or sometimes rediscovered. Gregory Razran (1971) reviewed more than 1500 empirical publications on learning and memory. His conclusion is that there must be a hierarchy of learning processes with several "superlevels", which can be related to the evolutionary stage where they first appear.

The lowest level is "reactive learning". It includes all animals and is exemplified by habituation and sensitization. On the next, "associative" superlevel (all vertebrates and some invertebrates) one finds classical and instrumental conditioning. The jump to higher vertebrates with its dramatic enlargement of brain and in particular with the development of hippocampus leads to a new class of learning processes. This "integrative learning" allows an animal to respond differently to a compound stimulus than to its components. Finally, in the next phase of phylogenesis the specifically-human, "symbolic" superlevel evolves. It deals with learning mediated by language processes as well as with learning about language and about learning itself. These functional mechanisms are unique to humans in using symbolic mediation, but reflect different stages of evolution, and all of them are at work in adult humans.

Though ultimate empirical proof or refutation of such "grand design" theories is a matter for future research, one cannot deny the quite different evolutionary origins of all the semi-autonomous "functional blocks", "memory systems", and "cognitive modules" investigated in contemporary psychology and neurosciences. Often these systems demonstrate different relationships to conscious and unconscious as well as to rational and prerational processing.

Obviously such cognitive mechanisms have to be built upon and to interact with hierarchical mechanisms of sensorimotor control. Half a century ago Nikolaj Bernstein (1947) attempted to give the first and as it seems still one of the most coherent pictures of such mechanisms. He differentiated four levels involved in control of an organism's movements: 1. Paleokinetic regulation, 2. Synergies, 3. Spatial field, 4. Object actions. Bernstein also supposed that "one or more" levels of "symbolic coordinations" may be localized "above" the level of object actions, which he identified with parietal and inferotemporal cortex.

Many of these early intuitions have gained empirical support in recent years. For example, grasping movements show a dependence of finer object-adjusted hand movements (action-level coordinations, according to Bernstein) on the global translatory motion of the arm (Bernstein's level of spatial field) (see Paillard 1987). A similar distinction is that between processing of spatial location and identification of individual properties of objects (see Section 5.3) necessary for their use and manipulation (Velichkovsky 1982;

Bridgeman 1992b). In both cases a coordination between neurological levels is required.

Even more important are some similarities to data on organization of human memory. From the time of the founding fathers of psychology (Ebbinghaus 1885) this field has been dominated by studies of what could now be called "explicit memory tests": intentional memorization of material for later recollection or recognition. Behaviorists (Thorndike 1913) discovered that memory for sensorimotor skills (swimming, biking, printing etc.) shows properties completely different from such intentional memory for verbal or generally symbolic information. First, it has an extremely enduring character and can be "reactivated" years after the initial learning. Second, this reactivation may be a surprise for the person. Third, to become reactivated memories of sensorimotor coordinations have to be supported "from outside", by some similar objective context – obviously you cannot discover that your skills to ride a bike are still in a good shape after a pause of several years if actually you are lying in bed or driving a car: one needs something like a bike for the discovery. This "object-driven", and not "concept-driven" character of motor automatisms explains why they may be grotesquely inadequate when triggered by a similar but inappropriate context. In other words, automatisms are if not necessarily prerational then often irrational, being at the heart of many action slips and mistakes (Norman & Shallice 1989).

The same dynamics is typical for the large class of memory effects discovered in the domain of perceptual and cognitive processing. In one of the first studies of this type Paul Kolers (see, e.g., Kolers 1975) investigated learning to read transformed texts. When one year after the learning subjects were suddenly invited to the lab for further exercises in reading the transformed texts they all demonstrated in this implicit memory test almost perfect saving of the reading skill achieved one year before. The effect is not limited to orthographic effects in reading of familiar words: important is to present subjects with more or less complicated patterns and not words per se. From all of traditional explicit memory tests it is recognition, especially recognition memory for complex ecologically relevant material, which comes closest to implicit memory, because it is massively supported "directly" from outside by the objective situation to be recognized. It is enough to say that – in contrast to the severe limitations of short-term memory – nobody yet knows what the limitations of recognition memory for meaningful pictures might be: in one study subjects were presented with 10,000 pictures, such as slides with urban or rural views, and one week later they could still correctly recognize about 80% of them (Standing 1973). There are data showing that implicit memory as well as recognition can be preserved if at the learning phase there was no attention or conscious control of the processing. Moreover, these effects remain relatively intact even in patients with amnesia, i.e. in people without memory in the traditional sense of the word (see, e.g., Hirst 1993).

Conceptualization of such data proceeds in terms of "memory systems" (Schacter & Tulving 1994) and "levels of processing" (Craik & Lockhart 1972). The discussion is mainly how many levels/systems there are, and whether they are levels of memory or perhaps parts of larger functional systems. A robust differentiation verified in hundreds of studies is between perceptual processing leading to perceptual priming and other implicit memory effects on one hand, and categorical processing leading to free recall and to semantic categorization, i.e., to relatively higher-level phenomena on the other. The latter seem to be situated more in the realm of conscious and rational processes. This simple picture is complicated by the flexibility of relationships between possible perceptual and cognitive factors in the course of learning. Thus, for instance, growing expertise in domains as different as chess, medicine, or machine construction can be accompanied by automatization of declarative knowledge and building up – on the lower levels of perceptual processing – of specialized automatisms. An expert therefore can literally see and hear relevant aspects of a meaningful situation (Velichkovsky 1994)

It would be also premature to conclude that the differentiation of perceptual and conceptual processing is the dividing line between rational and prerational intelligence. In fact, such relatively low-level effects as perceptual priming are not completely decoupled from rationality, but this is a case of a relatively "narrow-minded" rationality of "object-oriented" perception and action. For instance, an upside-down orientation or distortion in connectivity of naturalistic slides leads to an almost complete disappearance of the above-mentioned powerful effects of visual recognition memory (Velichkovsky 1982). In a provocative study, Cooper and Schacter (1992) demonstrated that in order to produce implicit memory effects, visual patterns in priming experiments must be two-dimensional depictions of three-dimensionally realizable ("possible") solid objects.

Returning to the higher symbolic coordinations one should mention a very special status of person-related cognition and memories (autobiographical memory, meta-memory). There are first of all robust neuropsychological dissociations of person-related cognitive processing from semantic memory. Thus, Tulving et al. (1991) demonstrated that patients could have intact semantic memory and even acquire new information, but their autobiographical memory may be severely damaged. In another study (see Velichkovsky 1999) as many as 30 different implicit and explicit memory tasks were compared in a common framework of the same experimental design. With one exception (free recall), metacognitive orientation to the material in the learning phase ("Evaluate the importance of these words to you now or in the future") led to significantly better results in unexpected implicit or explicit memory testing than all other forms of orientation, including instruction for intentional memorization ("Try to remember these words as well as possible for a future memory testing"). Metacognition is related to prefrontal, phylogenetically recent structures of cortex. Also in ontogenesis this is a late achievement (Kolband & Wishaw 1990). Is metacognition prerational, rational or perhaps postrational? In any case it is rather subjective, reflecting personal accents in evaluation of situations and in planning future activities.

Taking into account the earlier schema by Bernstein, with the addition of the results of the last decades, one can propose something like the following list of hierarchically evolving mechanisms leading to the human form of conscious rational decision-making (Velichkovsky 1990):

- *Level A. Paleokinetic Regulations.* Bernstein (1947) called this the "rubrospinal" level, having in mind the lowest structures of the spinal cord and brain stem regulating tonus, paleovestibular reflexes and basic defensive responses. The awareness of functioning is reduced to Head's (1920) protopathic sensitivity, so diffuse and lacking in local signs that even the term "sensation" seems to be too intellectual in this case.

- *Level B. Synergies.* Due to involvement of new neurological structures – the "thalamo-pallidar system" according to Bernstein (1947) – the regulation of an organism's movements as a whole is now possible: it becomes a "locomotory machine". The specializations of this level are rhythmic and cyclic patterns of motion underlying all forms of locomotion. Possibilities of awareness are limited by proprioceptive and haptic sensations.

- *Level C. Spatial Field.* A new spiral of evolution adds exteroception with the striatum and primary stimulotopically organized areas of the cortex as the control instances. This opens external space and makes possible one-time extemporaneous goal-directed motions in the near environment. The corresponding subjective experience is that of a stable voluminous surrounding filled with localized but only globally sketched objects.

- *Level D. Object Actions.* The next round of evolution leads to the building of secondary areas of neocortex that permit detailed perception and object-adjusted manipulations. Individualized objects affording some but not other actions come to the focus of attention. Formation and tuning of higher-order sensory-motor and perceptual skills is supported by a huge memory of the procedural type. Phenomenal experience is the perceptual image.

- *Level E. Conceptual Structures.* Supramodal associative cortices provide the highest integration of various modalities, supporting the ability to identify objects and events as members of generic classes. Development of language and human culture fostered this ability and led to formation of powerful declarative-procedural mechanisms of symbolic representation of the world. Common consciousness is the dominating modus of awareness at this level.
- *Level F. Metacognitive Coordinations.* In advanced stages, changes in conceptual structures result not only from accretion of experience but also from experimentation with ontological parameters of knowledge. Necessary support for this "personal view of the world" is provided by those parts of the executive areas of the vertebrate brain that show excessive growth in anthropogenesis, i.e. in the first line prefrontal neocortex. Coordinations of this level make possible personal and interpersonal reference, reflective consciousness and productive imagination.

Many of these levels of cognitive organization can be connected with specific stages of human ontogenesis. For example, manifestations of the theory of mind (Perner 1991) at about the age of 4 are clearly related to the prefrontal functions and metacognitive coordinations. Multi-level organization is also the key to an understanding of the numerous discontinuities and reorganizations of thinking in the course of cognitive development (Piaget 1983). It also can explain the apparently instant evolution of language which, as we have seen above, hardly could be possible without metacognitive and pragmatic influences on the preexisting lower-level mechanisms (see Section 5). The emerging picture is then rather of functional systems that are broader in their specialization than the function of memory. It is perfectly possible that emotion and motivation could also be integrated into the schema of the evolutionary developing mechanisms. In a clear way this idea was expressed by Heinz Heckhausen:

"At the lowest level we find the automatic reaction of the Autonomic Nervous System, the Endocrine and Immune Systems. Above this level there are prewired movement patterns for specific innate modes of behavior. Then follow primary drives which counterbalance the disturbances of the body's household. Above these we find acquired needs which are derived from the primary drives, but have become independent. Then there are all those primary affects as happiness, grief, fear, anger, surprise, and disgust that influence our experience And only then, on top of all this, do we find the higher, that is to say, social and cultural motives from which most of our wishes spring, provided that the lower systems do not happen to be busy managing homeostatic crises in the organism" (1985, p. 5).

Psychophysiological mechanisms of cognition, learning and memory demonstrate as we have seen profound differences along the vertical, phylogenetic dimension of their functioning. In evolution each level seems to emerge from lower levels. Then it starts to dominate these lower levels. Higher and lower levels coexist in phylogenetically later species.

The notion of "prerational intelligence" could be used of course in connection with the low to middle-level basis of everyday human intelligence, deeply rooted in object-oriented sensorimotor activity and perception. Corresponding forms of cognition and learning can be found also in experimental investigation of animals such as monkeys or even rats. Investigation in artificial neural networks, conducted in recent years, makes it possible that these forms of intelligence will be reconstructed in the new generations of artificial agents. Thus, it seems that when Artificial Intelligence started to build everything from the symbolic top, the whole construction had in a sense an upside-down orientation. Perhaps we are now at the promising point when this interdisciplinary endeavor could be put on its feet again.

5. THE DRIVING OF ACTION. THE PSYCHOPHYSIOLOGICAL APPROACH
(Jacques Paillard)

The explicit ambition of contemporary robotics is to explain how higher motor control and rational thinking might progessively emerge from a coordinated combination of basic sensori-motor devices that allow autonomous systems to survive in a given ecological niche. Indeed, we have seen in Section 2 how an "operative" intelligence based on constraining logical rules might progressively emerge in children's development from what Piaget has termed the "practical" intelligence of the early sensori-motor stages. Moreover, biologists believe that evolutionary pressures generally preserve basic neural mechanisms that have served to solve adaptation or survival problems in antecedent forms, but these mechanisms have to be restructured. Therefore, a coexistence and cooperation of a mental apparatus with a basic sensori-motor machinery is obviously required in the mature human brain (Paillard 1994).

Facing the impressive development of contemporary neurobiology, the challenge of exploring how behavioral and mental activities are tied to their biological roots and the various ways in which higher cognitive levels may adaptively control or supersede this basic machinery are to be considered (see also Paillard 2000).

5.1   *Voluntary vs. Automatic Motor Control*

The idea of anticipatory mental events, specifically associated with the voluntary initiation of goal-directed activity, was developed early in philosophical and psychological thinking. Buchanan (1812) put forward the basic concept of "an idea of action to be performed" as constituting, together with the "desire of performing" and "its execution" the three principal processes and essential components of a process of "volition". A profuse psychological literature has since been devoted to the elaboration of these concepts (reviewed in Kimble & Permuter 1970).

The dissociation between "voluntarily-driven" and "automatically-driven" actions was first established by neurologists (reviewed in Paillard 1982). Ideo-motor apraxia, first described by Liepmann, involves specifically the impairment of some transactional activity linking an intact "representation of the act" (a "motor image") with an existing "kinetic formula" in a patient unable to initiate a familiar act (making the sign of the cross, for example) when asked to do so, but automatically triggered by the appropriate context (entering a church, for example) to perform the same action. The reverse can be observed in patients deprived of their automatisms and thereby forced to rely on voluntarily controlled activities for all of their behavior.

Psychologists classically consider action-outcome representations as contributing to increasing the operating range of higher motor control and to incorporating novelty into the behavioral repertoire. It is a long-standing question whether these cognitive states, often called "goal or motor images" and even "kinesthetic engrams" by early neurologists, have a role in the control of behavior that is different from that of other motivational states that automatically trigger specific behaviors. The beginning of an answer is fortunately provided by the recent introduction of non-invasive neuro-imagery techniques in Neurology that allow the analysis of local metabolic activities during brain activities in man. Studies using cerebral blood flow measurement in man (Roland et al. 1980) showed that specific cortical areas (the supplementary motor areas) are activated during the initiation of voluntary movement but are not involved for the maintenance of a fixed posture or during the performance of automatic movement, thus supporting the idea of an anatomical segregation between neural structures underpinning voluntary and automatic control of motor activities.

Moreover, the increasing development of neuro-imagery techniques provides new ways to study motor activity in man in order to assess the contributions of various brain areas to neural events that precede the executive processes and are specifically associated with the planning and programming stages of movement production (review in Jeannerod 1994).

Extended practice of learned motor skills seems to produce a transition of control from a voluntary to an automatic form, thereby setting up a habit. A habit can be viewed as an action that, when triggered in a given situation, we perform automatically without deliberation or explicit reference to the goal of that action. For psychologists this transition would probably constitute a good example of what roboticists might consider as a transition from a "rational" to a "prerational" way of controlling behavior, or in other words between behavior "actively" or "passively" controlled by "intelligent processes". The distinction is also parallel to the "explicit" and "implicit" modes of memory theory (Parkin & Streete 1988). Neurophysiological studies support a two-stage model of motor skill acquisition (review in Paillard 1986) with a contribution of cortico-striatal loops in the early cognitive phase of motor learning, followed by a progressive transfer of automatized control to cerebellar circuitry. A beginning driver, for instance, must think about each motion. With experience, the activities of driving become so automatic that the driver can talk, eat or daydream while driving.

Figure 1, derived from pathological and neurophysiological data, shows the present neurobiological view of how the stream of information flows through various hierarchically organized levels of processing in the central nervous system. The various operations of intent, planning, programming and executing are serially articulated in the parallel activation and tuning of different internal loops. This kind of analysis guides experimental approaches in the field of animal and human motor control, and could provide interesting suggestions for the design of robots.

5.2 Reactive vs. Predictive Mode of Controlling Behavior

Converging evidence (reviewed in Goldberg 1985) from neuroanatomic studies, unit recording in behaving monkeys, movement-related cortical field potential and regional blood-flow studies has recently suggested a segregation between neural mechanisms subserving two basic modes, reactive and predictive, for generating and controlling behavioral activities.

Indeed, we can expect natural selection to shape the decision-making mechanisms of animals so that the basic sensorimotor repertoire of behavior sequences tends to be optimally adapted to the current situation in order to survive in a rather predictable environment, yet retain the flexibility necessary to assimilate and accommodate a reasonable range of unpredictable events. On the one hand, both the built-in mechanisms with their innate releasers identified by ethologists and the acquired habits incorporated in the repertoire of automatisms stored in the "dispositional memory" are "stimulus-driven" or more precisely "context-driven", allowing quick auto-

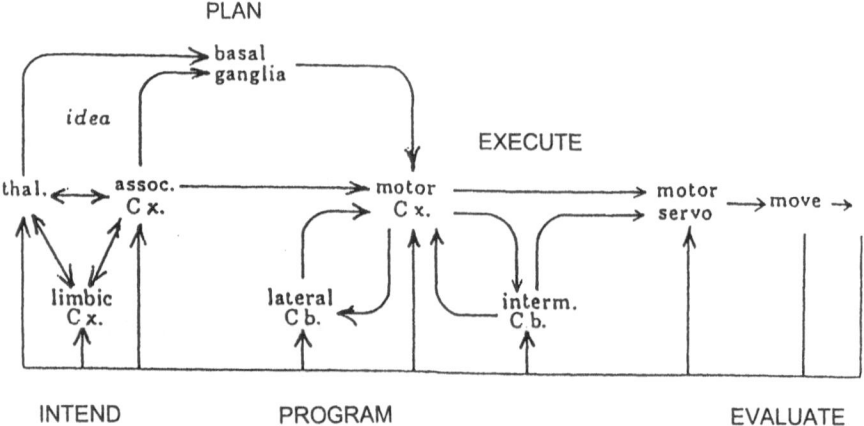

Figure 1: This scheme characterizes a number of types of activities and fits well both with clinical and behavioral data derived from lesion studies and with functional data derived from metabolic neuroimaging techniques (Paillard 1982).

matic triggering of adaptive reactions with low attentional demand and thus freeing cognitive awareness, with its limited capacity, for other supervisory operations. This is reactive control

On the other hand, the capacity to supersede such basic automatic reactivity characterizes the emergence of a predictive control of behavior that is oriented by prediction and cognitive anticipation of the consequences of the planned actions. The development of this capacity presupposes the ability to inhibit the primary stimulus-driven reaction and to escape the "tyranny" of the stimulus (Oakley 1985).

The anatomical distinction between a dorso-mesial and a ventro-lateral control system, converging on the primary motor areas but differently modulating the descending motor commands, corresponds precisely to the required mechanism for a dissociation between a reactive and a predictive mode of generating motor actions. Studies of the dorso-medial system emphasize the role of a supplementary motor area rostral to the primary motor cortical area. Experimental evidence suggests that this structure plays an important role in the generation of intention-to-act and in the specification and elaboration of planned action through its mediation between the primary motor cortex and the medial limbic cortex (classically considered as playing a major role in the selection of motivation-driven behavior). In contrast, the ventrolateral premotor area mostly receives exteroceptive multimodal information about the extra-corporeal environment and, therefore, mediates "stimulus-driven" modulation of behavior.

Moreover, neuroanatomical data derived from modern labeling techniques have established that two main premotor pathways for controlling executive commands in the motor areas correspond to a segregation in the cortical projections of the two main cortico-subcortical systems that modulate the activity of the premotor mechanisms.

On the one hand, the basal-ganglia loops project chiefly to the "predictive" dorso-mesial prefrontal system, suggesting its role is to process premotor events associated with planning and selecting the action and to sequence events appropriately for achieving a planned goal. Classically, impairment of this system leaves the organism able to drive and guide action only in a "reactive mode", based on visual or other sensory input.

On the other hand, the cerebellar loops project mainly to the ventro-lateral system and intervene mainly to regulate ongoing movement in accordance with changing postural states and to cope with unexpected changes in environmental constraints. They also contribute to building up an internal model of the stable and predictable dynamic constraints encountered by the executive mechanisms, which allows the automatic triggering of feedforward commands and thus alleviates the time lag of on-line feedback correction. One of us (Paillard) has studied a patient massively deprived of proprioceptive afferents from her body and hence of the availability of this type of cerebellar regulation. Her goal-directed motor behavior is entirely dependent on her highly attentionally demanding ability to drive it cognitively in a predictive mode.

Such a functional dissociation between reactive and predictive modes of generating behavior, together with its neurophysiological support, offers a coherent new frame for analysis of clinical and behavioral data. These would probably appear more appealing to psychologists and operationally more rewarding for empirical research than the rather ambiguous theoretical distinction between "prerational" and "rational" behavior proposed by robotics.

5.3 Sensorimotor vs. Cognitive Processing

The swinging of the pendulum from a hard behaviorism, restricted to a re-flexologic approach to behavior, to the excess of a neo-mentalism, rejecting biological knowledge as unnecessary for understanding psychological functions, has characterized Psychology during the last fifty years. This swing contributes to the gap between neuro-behavioral and cognitive approaches to brain function.

Neuroscientists, who are understandably concerned with the tractability of their experimental models, traditionally follow a bottom-up approach. For

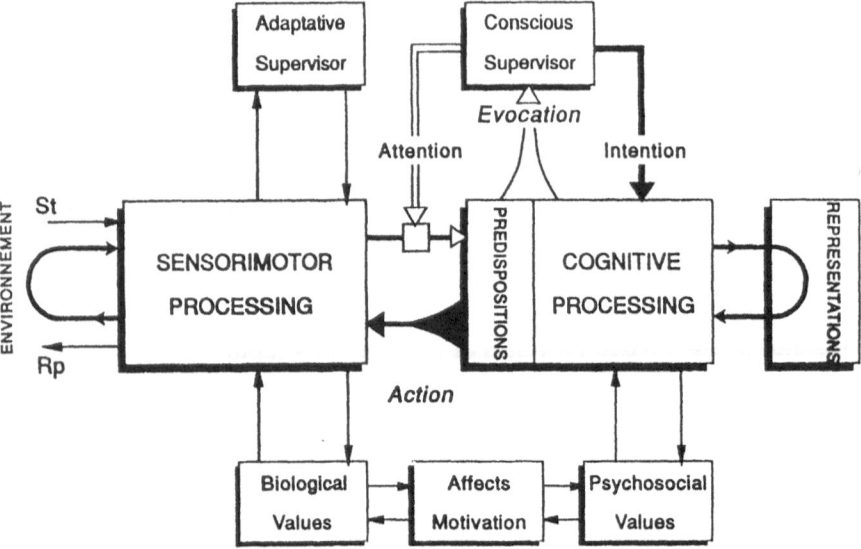

Figure 2: Via external loops, the sensorimotor compartment directly relates sensory information to motor activities which, in turn, contribute to transforming the sensory field. The
concept of sensorimotor dialogues implies that the environment and the organism are mutually and alternately sources of questions and suppliers of responses (Paillard 1999, Fig. 4).

a long time, this approach has imposed on their experimental procedures a
neutralization of higher levels in order to allow the study of basic mechanisms in isolation.

Therefore, psychologists generally thought that neuroscientists overlooked the powerful modifiability of basic sensorimotor mechanisms by higher
mental processes and, as we ascend the phylogenetic scale, their increasing
submission to inference and expectancy. Cognitivists, for their part, were
comforted by the success of the linguistic and computational approach of
A.I. But in the opinion of neuroscientists, cognitivists – in their desire to further the autonomous development of their discipline – ignored the neural implementation of mental operations and overlooked the role of the hard wiring
of the nervous system and of sensorimotor routines that are the mandatory
link between the physical world and its neural representation in the brain.
The vigourous progression of neurosciences has recently opened the way
for a renewed interdisciplinary approach to brain function and behavioral
studies.

As part of this approach the basic segregation between a multilevel organization of sensorimotor and cognitive operations now demands recognition
in both communities. Such a dual system of information processing is presented in Figure 2 (from Paillard 1999).

The cognitive processing compartment handles the internal dialogue between the cognitive apparatus and stored mental representations of the physical environment, under the supervision of a conscious evaluator and the monitoring of attentional and intentional processes.

The schema obviously requires input and output routes. The cognitive compartment has an input channel, which includes attentional gating, and two output channels. The motor-oriented output channel has a repertoire of learned motor schemata and strategies that has long been regarded as the natural output of the cognitive apparatus and probably the only one that can process dynamic cognitive processes.

The evocation channel is the route by which information enters conscious experience in the form of mental images or re-representation of propositionally coded cognitive data. Access to the evocation channel is through predisposed built-in perceptual schemata. Schemata are here envisioned as high-order tuned neural networks for translation both of the evocative level of sensory inputs and of representational knowledge. They also allow transformation from goal images to intended acts through a repertoire of motor plans, and they mediate a direct mobilization of motor habits permanently stored in dispositonal memories (Thomas 1984). An autonomous supervisor, adjunct to the sensorimotor compartment, allows parametric adjustments to be set automatically in relation to biologically relevant values or to preserve kinetic invariance in accordance with postural requirements of body balance.

The basic affordance of the physical world (in a Gibsonian sense) is reflected in the aggregation of neuronal ensembles in integrated sensorimotor units. The regularities and covariant features of reafferent information that are recorded in the central stores of dispositional memory constitute the basic repertoire of learned perceptual schemata. These characterize the input stage of sensorimotor units; innate or learned motor schemata characterize their output stage. Interactions between an organism and its environment exhibit basic inbuilt "intelligent" designs to explore and manipulate information. These interactions enrich the stored representations of the organism's knowledge of both its own body and of external objects and events.

In contrast, cognitive processes refer to computational transactions that incorporate these stored representations in some kind of internal dialogue. In this sense, representations allow the organism to step back from the immediate constraints of on-line sensorimotor dialogues and bestow on the brain its increasing capacity to predict and control perceptual and motor activities. In a simple example of this multilevel control, a spinal reflex withdraws the hand from a hot stove before the brain even finds out about the event. At a higher level, the same event will cause the subject to be more careful around hot stoves in the future. No single level can provide both functions.

Such a view fits with that of animal robotics chiefly interested in robots endowed with those basic autonomous adaptive properties that the sensorimotor compartment bestows on living systems. Indeed such biologically founded autonomy justifies the claim of roboticists that they do not need to explain the "intelligent" behavior of purely sensorimotor organisms. Any requirements for higher level psychological notions such as attention, intention, consciousness, images of goal and the like are judged to be unnecessary anthropomorphisms. Obviously, this view is consistent with the study of physiological functions.

However, the existence of an animal subsymbolic cognition is well recognized (Oakley 1985). It provides grounds for understanding how a cognitive agency, generated primarily by the evolutionary elaboration of sensorimotor machinery reactively driven by biological values, can cohabit, interact and cooperate with such a basic interface with its physical and social environment, and extend its power of prediction and autonomous control. The concept of "cognitive penetrability" (Pylyshyn, 1981) now pervades the neuropsychological approaches of disordered higher brain functions. In line with developmental studies mentioned above, we need further study in animals and humans of how sensorimotor dialogues might contribute to shaping and tuning cognitive operations in the developing and working brain.

6. CONCLUSIONS

Until recently the only possible approach to high-level cognition was a computational theory of mind based primarily on the manipulation of symbolic representations, as promoted by AI approaches. Alternative approaches, variously termed "Artificial life", "Adaptive systems", and "Autonomous Robots", have made their appearance and are more appropriate for the study of lower level expression of intelligent behavior.

These new approaches reflect a pragmatic resurgence of considering "cognition" as an emergent phenomenon, proper to autonomous systems and derived from the sensorimotor dialogues that tie the organism to its niche and allow its survival. In such an autoreferential system, function is assumed to emerge as a consequence of a value-driven association between an input state and a motor output.

In the mature organism there is a coexistence and cooperation of a cognitive apparatus with a basic sensorimotor machinery and the various ways in which higher cognitive levels may adaptively control or supersede this machinery, taking into account intrinsic constraints that set definite bounds to their operating range.

The challenge is seriously worth considering in its possible consequences for the development of our disciplines. First, it stimulates a reappraisal in

Psychology of psychobiological theories about how behavioral and mental activities are phylogenetically and ontogenetically tied to their biological roots. Second, it induces in the modellers of autonomous systems an increasing awareness of the requirement of biological plausibility.

New tools emerge for the experimental study of mental imagery and cognitive operations in man (Roland et al. 1980). Potential resources include neuro-imaging techniques that will offer in the near future new possibilities for empirical validation of what are still considered speculative hypotheses.

Further study in animals and humans, in line with developmental studies, reviewed above, of the way in which sensorimotor dialogues might contribute to shape and tune the cognitive operations in the developing and working brain, will provide a fertile arena for interdisciplinary exchanges. Biological, behavioral, psychological and computational approaches will each make a contribution.

University of California, Santa Cruz, CA

**Université de Genève, Switzerland*

***Centre National de la Recherche Scientifique – Neurobiologie et Mouvement, Marseilles, France*

****Technische Universität Dresden, Germany*

REFERENCES

Arbib, M.A. (1972). *The metaphoric brain. An introduction to cybernetics as artificial intelligence and brain theory.* New York: Wiley-Interscience.

Baars, B.J. (1988). *A cognitive theory of consciousness.* New York: Cambridge University Press.

Baddeley, A.D. (1986). *Working memory.* Oxford, UK: Oxford University Press.

Bernstein, N.A. (1947). *O postrojenii dvizhenij [On construction of movements].* Moscow: Medigiz.

Bickerton, D. (1983, July). Creole languages. *Scientific American* **249**, 108–115.

Bickerton, D. (1984). The language to consciousness. In E.R. John (ed.), *Machinery of the mind* (pp. 128–156). Cambridge, MA: Birkhäuser.

Bidell, T. (1988). Vygotsky, Piaget, and the dialectic of development. *Human Development* **31**, 329-348.

Binet, A., & T. Simon (1916). *The development of intelligence in children.* Trans. by Elizabeth S. Kite. Baltimore: Williams & Wilkins.

Brentano, F.C. (1874). *Psychologie vom empirischen Standpunkt.* Leipzig: Duncker und Humblot.

Bridgeman, B. (1986). Relations between the physiology of attention and the physiology of consciousness. *Psychological Research* **48**, 259–266.

Bridgeman, B. (1988). *The biology of behavior and mind. Ch. 14: Consciousness and high-level control.* New York: Wiley.

Bridgeman, B. (1992a). Conscious versus unconscious processes: The case of vision. *Theory and Psychology (Sage)* **2**(1), 73–88.

Bridgeman, B. (1992b) The evolution of consciousness and language. *Psycoloquy.***92***. Consciousness.***1.

Brooks, R.A.(1991). New approaches to robotics. *Science* **253**, 1227–1232

Buchanan, J. (1812). *The philosophy of human nature*. Richmond, Kentucky: Grimes.

Cellérier, G. (1992). Le constructivisme génétique d'aujourd'hui. In B. Inhelder & G. Cellérier (eds.), *Le cheminement des découvertes de l'enfant* (pp. 217–253). Neuchâtel, Switzerland, and Paris: Delachaux et Niestlé.

Cooper, L.A., & D.L. Schacter (1992). Dissociations between structural and episodic representations of visual objects. *Current Directions in Psychological Science* **1**, 141–146.

Craik, F., & R. Lockhart (1972). Levels of processing; A framework for memory research. *Journal of Verbal Learning and Verbal Behavior* **11**, 671–684.

Dean, A.L. (1994). Instinctual affective forces in the internalization process: Contribution of Hans Loewald. *Human Development* **37**, 42–57.

Ebbinghaus, H. (1985). *Über das Gedächtnis*. Leipzig, Germany: Duncker & Humblot.

Edelman, G.M. (1987). *Neural Darwinism*. New York: Basic Books.

Goldberg, G. (1985). Supplementary motor area structure and function: Review and hyptheses. *Behavioral and Brain Sciences* **8**, 567–616.

Head, H. (1920). *Studies in neurology*. Oxford, UK: Oxford University Press.

Heckhausen, H. (1985). *Wünschen – Wählen – Wollen*. In H. Heckhausen, P.M. Gollwitzer, & F.E. Weinert (eds.), *Jenseits des Rubikon: Der Wille in den Humanwissenschaften* (pp. 3–9). Berlin: Springer.

Heckhausen, H. (1991). *Motivation and action*. Berlin: Springer Verlag.

Hirst, W. (1993). On the nature of systems. In G. Harman (ed.), *Conceptions of the mind*. Hillsdale, NJ: Lawrence Erlbaum Assoc.

Inhelder, B., & G. Cellérier (1992). Le cheminement des découvertes de l'enfant. Lausanne, Switzerland: Delachaux et Niestle.

James, W. (1890 [1950]). *The principles of psychology*. New York: Dover Publications.

Jeannerod, M. (1994).The representing brain: Neural correlates of motor intention and imagery. *Behavioral and Brain Sciences* **17**, 187–245.

Kimble, G.A., & L.C. Perlmuter (1970). The problem of volition. *Psychological Reviews* **77**, 361–384.

Koehler, W. (1925). *The mentality of apes*. New York: Penguin.

Kolb, B., & I.Q. Wishaw (1990). *Fundamentals of human neuropsychology*. San Francisco, CA: W.H. Freeman.

Kolers, P. (1975). Specificity of operations in sentence recognition. *Cognitive Psychology* **7**, 289–366.

Luria, A.R. (1981). *Language and cognition*. Ed. and translated by J.V. Wertsch. New York: Wiley.

McFarland, D. (1989). The teleological imperative. In A. Montefiore & D. Noble (eds.), *Goals, no-goals and own-goals* (pp. 211–228). London: Unwin Hyman.

McFarland, D., & T. Bösser (1993). *Intelligent behavior in animals and robots.* A Bradford Book. Cambridge, MA: MIT Press.

Miller, G.A. (1956). The magical number seven plus or minus two: Some limits on our capacity for processing information. *Psychological Review* **63**, 81–97.

Miller, G.A., E.H. Galanter, & K.H. Pribram (1960). *Plans and the structure of behavior.* New York: Holt Rinehart & Winston.

Minsky, M.L. (1986). *The society of mind.* New York: Simon and Schuster.

Newell, A., & H.A. Simon (1972). *Human problem solving.* Englewood Cliffs, NJ: Prentice-Hall.

Norman, D.A., & T. Shallice (1980). *Attention to action: Willed and automatic control of behavior.* Technical Report 8006, Center for Human Information Processing. San Diego, CA: University of California.

Oakley, D.A. (1985). *Brain and mind.* New York: Methuen.

Paillard, J. (1982). Apraxia and the neurophysiology of motor control. *Philosophical Transactions of the Royal Society London* **B 298**, 111–134.

Paillard, J. (1986). Development and acquisition of motor skills: A challenging prospect for neuroscience. In H.T.A. Whitting & M.G. Wade (eds.), *Motor development in children: Aspects of coordination and control* (pp. 416–441). Den Haag, The Netherlands: Martinus Nijhoff.

Paillard, J. (1987). Cognitive versus sensorimotor encoding of spatial information. In P. Ellen & C. Thinus-Blanc (eds.), *Cognitive processes and spatial orientation in animal and man* (pp. 43–77). Dordrecht, The Netherlands: Martinus Nijhoff.

Paillard, J. (1994). L'intégration sensori-motrice et idéo-motrice. In M. Richelle, J. Requin, & M. Robert (eds.), *Traité de psychologie expérimentale* (pp. 925–961). Paris: Presses Univérsitaires de France.

Paillard, J. (1999). Motor determinants of a unified word perception. In G. Aschersleben, T. Bachmann, & J. Müsseler (eds.), *Cognitive contributions to the perception of spatial and temporal events* (pp. 95–111). Amsterdam: Elsevier.

Paillard, J. (2000). The neurobiological roots of rational thinking. In H. Cruse, H. Ritter, & J. Dean (eds.), *Prerational intelligence: Adaptive behavior and intelligent systems without symbols and logic,* Vol. 1 (pp. 343–355). Dordrecht, The Netherlands: Kluwer Academic Publishers.

Parkin, A.J., & S. Streete (1988). Implicit and explicit memory in young children and adults. *British Journal of Psychology* **79** (3), 361–369.

Perner, J. (1991). *Understanding the representational mind.* Cambridge, MA: MIT Press.

Piaget, J. (1952). *The origin of intelligence in children.* New York: Norton.

Piaget, J. (1983). Piaget's theory. In P.H. Mussen (ed.), *Handbook of child psychology, Vol. 1.* New York: Wiley & Sons.

Pylyshyn, Z.W. (1981). The imagery debate: Analogue media versus tacit knowledge. *Psychological Review* **88**, 16–45.

Razran, G. (1971). *Mind in evolution.* Boston, MA: Houghton-Mifflin.

Roland, P.E., B. Larsen, N.A. Larsen, & E. Skinhoj (1980). Supplementary motor area and other cortical areas in the organization of voluntary movements in man. *Journal of Neurophysiology* **43**, 118–136.

Ryle, G. (1949). *The concept of mind.* New York: Barnes and Noble.

Schacter, D.L., & E. Tulving (1954). *Memory systems.* Cambridge, MA: MIT Press.

Shallice, T. (1978). The dominant action system: An informal model. In K. Pope & J.E. Singer (eds.), *The stream of consciousness: Scientific investigation into the flow of conscious experience* (pp. 117–157). New York: Plenum.

Shimamura, A.P. (1994). The neuropsychology of metacognition. In J.J. Metacalfe & A.P. Shimamura (eds.), *Metacognition: Knowing about knowing.* Cambridge, MA: MIT Press.

Standing, L. (1973). Learning 10,000 pictures. *Quarterly Journal of Experimental Psychology* **25**, 207–222.

Stewart, J. (1994). The implications for understanding high-level cognition of a grounding in elementary adaptive systems. In P. Gausser & J.-D. Nicoud (eds.), *From perception to action* (pp. 312–317). Los Alamitos, CA: IEEE Computer Society Press.

Thomas, G.J. (1984). Memory: Time binding in organisms. In L.R. Squire & N. Butters (eds.), *Neuropsychology of memory* (pp. 374–384). London: Guilford Press.

Thorndike, E.L. (1913). *The psychology of learning.* New York: Teachers College.

Tulving, E., C.A. Hayman, & C. MacDonald (1991). Long-lasting perceptual priming and semantic learning in amnesia. *Journal of Experimental Psychology: Learning, Memory and Cognition* **17**, 595–617.

Velichkovsky, B.M. (1982). Visual cognition and its spatial-temporal context. In F. Klix, J. Hoffmann, & E. van der Meer (eds.), *Cognitive research in psychology* (pp. 47–58). Amsterdam: North Holland.

Velichkovsky, B.M. (1990). The vertical dimension of mental functioning. *Psychological Research* **52**, 282–289.

Velichkovsky, B.M. (1994). The levels endeavour in psychology and cognitive science. In P. Bertelson, P. Eelen, & G. d'Ydewalle (eds.), *International perspectives on psychological science: Leading themes* (pp. 143–158). Hove, UK, and Hillsdale, NJ: Lawrence Erlbaum Assoc.

Velichkovsky, B.M. (1999). From levels of processing to stratification of cognition. In B.H. Challis & B.M. Velichkovsky (eds.), *Stratification in cognition and consciousness* (pp. 203–226). Amsterdam: John Bejamins.

Vygotsky, L.S. (1962). *Thought and language.* Ed. and translated by E. Hanfmann and G. Vakar. Cambridge, MA: MIT Press.

Vygotsky, L.S. (1978). *Mind and society.* Cambridge, MA: Harvard University Press.

Wertsch, J.V. (1985). *Vygotsky and the social formation of mind.* Cambridge, MA: Harvard University Press.

HELGE RITTER

PRERATIONAL INTELLIGENCE FROM THE PERSPECTIVES OF ROBOTICS AND ENGINEERING

1. INTRODUCTION

Robots are a type of machine that we expect to solve tasks that are in many respects similar to those that confront animals and human beings. We want them to be able to perceive their environment through sensors that may include vision and touch, they should be able to move and to avoid collisions with obstacles, they should have manipulators that allow them to grasp and manipulate work pieces and, ideally, they should be able to cooperate with humans in a manner that is convenient – at least for us.

One of the scientific values of the field of robotics is that it teaches us what in addition to rational intelligence is required to act successfully in a real world in which the presence of objects – together with their properties – must first be derived from sensory signals and sensori-motor interaction with the environment. By trying to build technical artefacts that can perform some of the activities of animals or humans we are forced to synthesize a subset of the functions that nervous systems provide in order to extract stable entities from complex and highly variable spatio-temporal sensor patterns (Richards 1988; Cruse 1996). Since our own brain performs this task in such a superb and effortless way, we easily fall victim to the illusion that the objects that we perceive are clear, well delineated and highly "obvious". The true difficulty of the required processes only becomes apparent when we try to build robots for solving similar tasks.

Therefore, from the perspective of prerational intelligence, robotics has the status of an experimental science that can correct our distorted perception about the processes involved in intelligent behavior. By trying to build robots that can solve tasks in the real world we can explore the particular difficulties connected with the identification of objects, their classification, for instance into obstacles or work pieces, and the coordination of movements in order to achieve a particular goal (Brooks 1989).

Nowadays, the most widespread type of robot is the industrial robot. Its mechanical design usually resembles that of a large arm, carrying a tool or an end-effector. Typically, these robots still perform rather simple tasks, involving a rather stereotypical course of movements, such as spray painting or soldering a work piece that passes by on a conveyor belt. The highly

135

J. Dean et al. (eds.), Prerational Intelligence: Interdisciplinary Perspectives on the Behavior of Natural and Artificial Systems, 135–159.

predictable and structured environment in which such robots perform their job allows them to fulfil their task with a minimum of sensor use. This makes them rather different from animals or humans, which have to survive in a highly variable environment.

2. PLANNING AND WORLD MODELS

When any actions in such an environment are required, the first issue that may come to mind is the issue of *planning*. Therefore, in particular in artificial intelligence research, the issue of robot planning has received considerable attention (Brady et al. 1982; Lozano-Perez 1987; Lozano-Perez et al. 1989). Usually, one distinguishes at least two different levels of planning: planning at the task level, and planning at the lower level of trajectory formation. More recently, it also has become apparent that it may be necessary not only to plan movements and actions, but also to plan sensor use (Hager & Mintz 1991; Dario et al. 1992).

The initial planning approaches focused on largely idealized situations, such as the movements of a mobile robot in an environment that is fully known. This led to robot planning algorithms such as the "classical" STRIPS approach (Fikes et al. 1971) and its later refinements, in which the actions of the robot were described formally as a set of "operators" (for instance, "move one step left" or "move one step forward") that change a "world state" (which may be given by a set of logical predicates, which might specify coordinates, e.g. the position of the robot in a two-dimensional grid, with adjacent grid positions always one step apart). Planning then consisted in finding legal operator sequences that led to a path in state space that connects the given starting state to the final goal state. This task has to be solved under the important constraint that there may be unrealizable world states (e.g., the robot may not occupy the same position as an obstacle). Once the task has been formulated in this way, methods of logical deduction, in conjunction with search algorithms, can be applied to find an operator sequence (a "plan") that solves the given task under the given constraints (provided these admit a solution at all).

In principle, this is a very general approach, since the world state can (and usually has to) include many more variables than just the position of the robot. Together with the rules for how the operators change the world state, this constitutes what is usually called a "world model", which provides the planning process with an abstraction of the real situation.

For quite some time this framework was more or less tacitly adopted in the field of robotics (Agre & Chapman 1990). However, the experiences gained with that approach often turned out to be rather frustrating. Over time it became apparent that the conceptually attractive idea of a world model contains several assumptions which are rather strong and may be difficult to meet in practice: First, we have to ascertain that a world model can be made available at all, which may be rather difficult for many realistic situations; second, the world model must be sufficiently simple so that the planning can be done within reasonable time and memory bounds; and third, the computed plan, when carried out under real circumstances, should lead to a similar result as in the world model that was used for the planning.

Let us consider the implications of these assumptions in some more detail. In any real environment a world model can at best represent some major constraints. Almost all of the details are unknown and can only be estimated. One might argue that details should not matter too much. Unfortunately, this is wrong in many situations. When we grasp an object, it matters much whether its surface is slippery or rough, whether it is soft and deformable, or rigid and hard, whether it is light or heavy and so on. Usually, we cannot know the values of these object parameters beforehand. Correspondingly, our grips are hardly preplanned, but instead are sophisticated strategies to rapidly identify the missing properties to the necessary extent during the initial grip phases with corresponding, rapid adjustments of our movement thereafter (Johansson 1996). In fact, the control of even rather simple mechanical systems, such as an inverted pole, can not be preplanned unless our model and our computations were of infinite accuracy. Even the tiniest error in a precomputed plan will soon amplify and lead to total failure within a short timespan if we do not continuously correct our motions on the basis of continual sensory feedback.

Coming to the second point, the feasibility of planning in a given world model, we face new difficulties. One of them is the "curse of dimensionality", which means that there are usually more relevant variables than we would like to consider, leading to high dimensional representations with computationally very expensive search procedures. Even if this problem can be surmounted for a single robot, we are not yet done in many cases. In a more realistic situation, there will be other "agents" in the environment, and their actions will at best only be partly predictable. Therefore, we can only plan some gross elements of the overall action and a large part of the actual actions can only be filled in when the real situation unfolds and the activities of the other agents can be observed. Note that this difficulty cannot be circumvented by some "internal simulation": we simply do not have the information that would be required to predict the reactions of the other agents in our internal simulation!

But even if we had this information, the third point presents further obstacles. Imagine that the other agents were as simple as a wooden dice on a table, which is subject to very precisely known laws of Newtonian mechanics. While we then (quite) easily could run a very detailed simulation of the dice's "behavior", this would not be very useful at all. We intuitively know that the movement of a dice depends in a subtle way on numerous small disturbances that occur in its initial conditions and while it rolls over the table. Dynamical systems theory (Arrowsmith 1994) has made this mathematically precise and has shown that for systems of the type exemplified by the dice (or the atmosphere, to mention another famous example) tiny uncertainties in the initial conditions inevitably amplify over time to such an extent that the outcome some time later is described most adequately as a random quantity, notwithstanding that we know the precise laws and can set up a simulation of high numerical accuracy (Schuster 1988). This phenomenon of effective non-predictability of even rather simple systems has been known to mathematicians for a long time (and the relevant part of their theory has become widely known as "chaos theory" less than two decades ago). Its implications for robotics, and, more generally, for the idea of world models and planning in AI, however, have been appreciated only much more recently.

These difficulties do not mean that concepts like planning and symbolic reasoning are useless; however, they show us that to build a robot that can successfully cope with the demands of a "natural" environment we first have to consider issues that we would classify into the realm of prerational intelligence. Without solutions to these (apparently!) "simpler" issues logical deduction and reasoning – the hallmark activities of "rational intelligence" – have no material on which they could work (Brooks 1990).

3. PERCEPTION

As we have seen, the strengths of logical deduction and reasoning are bought at the expense of a rather severe restriction: they need as their starting point always a "clean" symbolic description of the posed task. It is the task of *perception* to derive a suitable description – to the extent possible – from the real world.

While the effortless functioning of our own perception makes us entirely unaware of the complexity of the underlying processes, their difficulty becomes highly apparent as we try to build technical devices that exhibit at least some sophisticated perception capabilities (Hurlbert & Poggio 1988; Ballard 1991).

Imagine, for instance, an assistance robot on a construction site, whose task is "just" to autonomously dig out large stones and carry them to some destination place. A human instructor may direct the robot by indicating the next stone and its destination place by a pointing gesture, he also might use language, or a combination of both.

In this case, the robot would have to use vision to recognize "his" human instructor among other workers that may be accidentally around. It would have to identify arm movements and to distinguish meaningful pointing gestures from other, non-task related arm and body motions (Littmann et al. 1996). When the robot has detected a pointing gesture (which in itself is already a highly complex vision problem (Nölker & Ritter 1998), particularly when detection has to work highly reliably under the broad range of lighting conditions on a construction site), it has to relate it with the local environment: is there something in the direction of the pointing gesture that resembles a stone? This is again a vision task, but with a target object that can vary in a broad range in shape and color; moreover, usually most of the stone will be covered by the ground and only a small part visible.

Therefore, an important part of the perceptual apparatus of a robot is a good vision system. Additional requirements are sensors to measure forces. For really dextrous manipulation of objects further capabilities, e.g., haptic discrimination of surface properties, would be desirable. Each of these demands easily gives rise to an entire research area (Brady 1989); in the following, we shall only consider the aspect of vision, since it probably is sufficient to highlight the eminent role of prerational processes to preprocess the inputs of our senses into a form that is suitable as a starting place for our rational intelligence.

Fig. 1 shows the picture of a simple scene. We immediately can recognize an object, distinguish its parts and judge their relative positions, as well as many further properties. A well-working vision system (like ours) has to extract these structures from the raw image data. To allow us to appreciate the huge difference between the original sensory input and what our vision system makes us perceive, we transform the camera input that gave rise to Fig. 1 into a representation to which our vision system is not well adapted: instead of using gray values for the representation of pixel intensities we simply represent the intensity pattern as a "landscape" whose height above a point (x, y) in the ground plane represents the image intensity at that location (Fig. 2).

While we routinely use this type of representation for the display of many kinds of data, the application of the same method to an image makes the image contents practically inaccessible to us. The reason is that we now have "disabled" the prerational processing abilities of our visual system by representing the image data in a "wrong" format (although it still contains the same information). We still can use our rational intelligence to infer where peaks and valleys in the landscape indicate features of objects, but

Figure 1: A gray level picture of a wooden toy object.

we see that this is a cumbersome procedure that deprives us of most of the information that the image contains.

The first attempt to create artificial vision for robots go back into the sixties where the pioneering works of Roberts (1965) and later of Waltz (1972) showed how to recognize the shape and pose of polygonal objects from their images. These initial approaches and much of the work they triggered focused on the extraction of edge elements, their connection into longer line segments and the use of additional geometric knowledge to ultimately derive a 3-dimensional description of the objects in a scene. While the initial approaches only worked under carefully controlled lighting conditions and good contrast, subsequent work managed to exploit more cues about object shape, such as shading, texture (Horn 1977), stereo disparity or specular highlights, and thereby to achieve higher robustness (Poggio et al. 1985). A comprehensive and systematic framework for how to achieve this integration was given by Marr's famous work, which presented a complete theory of how to derive from the raw image data a rich representation (called the "$2\frac{1}{2}$D-sketch") which then would form the basis for the final step to a full 3D-reconstruction (Marr 1982). While this was a significant development, the idea of reconstructing a 3D-scene description was fully in line with the then still dominating assumption that the construction of an explicit world

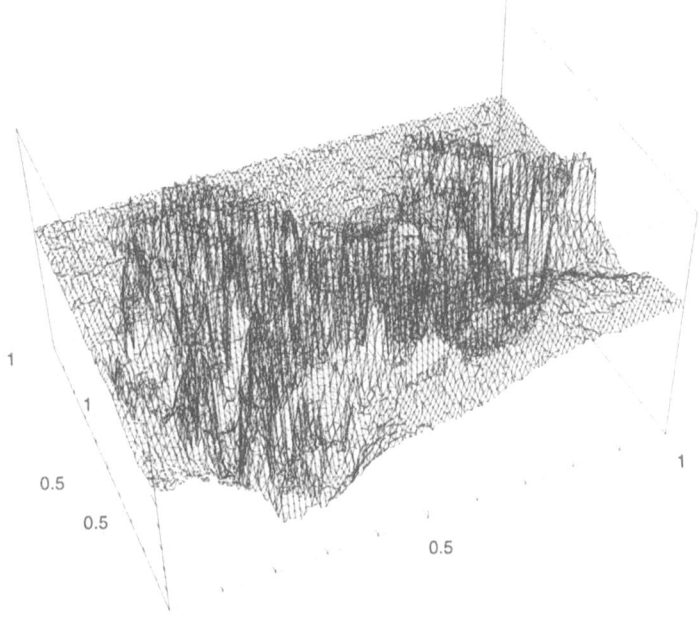

Figure 2: The same information, but now with gray values depicted as heights of a land-scape. The task of a vision system is extract information about the presence of objects and their properties from sensor data of this kind.

model was the right way to go. Only later, by studying animal and human perception of 3D shape for unfamiliar objects (Bülthoff & Edelman 1992; Logothetis et al. 1994), it became apparent that the necessary information for a full 3D reconstruction simply may be unavailable in many situations and that our impression that we can "see" the three-dimensional shape of an object results from our familiarity (and thus our knowledge of their three-dimensional shape) with most of the objects that we already recognize from their two-dimensional views (Cutzu & Edelman 1994).

This observation has led to approaches to vision that emphasize a view-based recognition of objects without necessarily producing a 3D reconstruction of their shape (Murase & Nayar 1995; Nayar & Poggio 1996; Edel-man et al. 1992; Weng et al. 1993; Heidemann & Ritter 1996). It turned out that this – in some way more modest approach – leads to find algo-rithms that can be very fast by performing a "holistic" object recognition through a non-linear mapping from the image intensity pattern through a small number of intermediate stages into a set of object descriptors (in the simplest case, a number of discrete object classes). The intermediate stages are often highly parallel and the entire algorithm can often very naturally be implemented as a layered network of artificial neurons. The latter feature entailed another shift towards "prerational intelligence": while in the classical vision approaches most processing steps were attempted based on explicit mathematical considerations about the nature of the imaging process, the transformations in a neural network result from a large number of weights whose optimal values only in exceptional cases can be derived by some direct analytical means. Usually, suitable values have to be found from learning approaches, using a training set of images for which the desired output of the recognition network is known. By means of suitable learning rules, the neural connections can then be iteratively adjusted so that the actual output of the network gradually approaches the desired output (for more details on these methods, see the chapters *Mathematical Perspectives on Prerational Intelligence* and *Computer Science Perspectives on Prerational Intelligence*). The result of this type of approach then is a network whose "visual knowledge" is no longer encoded in humanly interpretable rules[1] that are applied to transform the intermediate representations, but instead in a large set of weight values that encode the visual knowledge in a "distributed" and usually rather unintelligible way. A further, and rather attractive, feature is that this makes it possible to extract visual knowledge from examples, i.e., there is no longer any need to have the required knowledge encoded in an explicit form. This is particularly useful for domains such as vision, where we all are highly skilled "experts", but with almost no explicit knowledge concerning to which "rules" we use to perform our skill.

Even when the robot's sensory processing system has managed to solve all these tasks, only a small part of a "clean task description" has been obtained: there are still many unknowns, such as the shape and extent of the stone under the ground, the mechanical properties of the soil (influenced by weather conditions, such as rain!) and so on. Therefore, even now there is no basis for extensive planning; returning to our example, the robot must instead begin to explore the stone by starting to dig around it in the ground. The response will be a temporal pattern of reaction forces, depending in a complex way on the robot's own actuator movements and the properties of the ground and the stone. The situation is familiar to everyone who has ever used a spade to dig out a rock. In such a situation, we hardly ever have a problem in deciding how to move our spade: we are not at all concerned with our rational intelligence, but instead use our prerational intelligence for an effortless and largely unconscious evaluation of the perceived resistance forces to gain an impression about the true size of the stone and to direct our digging.

Whenever it comes to an interaction with the real world, it is tasks like this that are of primary concern. Sometimes, their solution leads to a situation that then may call for logical deduction and systematic planning. But in many cases, even that does not occur and the whole activity takes place entirely in the realm of prerational intelligence.

4. KINEMATICS AND DYNAMICS

This does not mean that at these lower levels there is no planning involved at all. However, it seems that the planning that occurs at the "prerational level" is not so much concerned with symbolic and very general planning strategies, but is much better described as a continuous optimization process, often exploiting rather specific properties of the particular task.

Actually, despite their rather predictable environment, even industrial robots face some of these "prerational planning" tasks for their actions: they must be able to move their arms along a prescribed trajectory, and they are expected to do so at a fast rate and usually with a high degree of accuracy. Good solutions to this task are much less trivial than it might appear. Like with most of our body functions, we hardly know "how" we ourselves move our arms. In fact, moving the human arm involves the coordination of considerably more than the seven degrees of freedom of the arm itself: the reaction forces caused at the shoulder call for adjustments in numerous other muscles in our trunk in order to stabilize our posture. All these adjustments are coupled in a rather complicated way, are fully automatic and their correct generation constitutes a major feat of the prerational intelligence that is built into our motor system (Flash & Hogan 1985; Massone & Bizzi 1989).

Finding a good trajectory for an industrial robot requires us to analyse these couplings at least for an arm with six degrees of freedom (the standard design of an industrial manipulator; six independent degrees of freedom is the minimal number required to allow arbitrary object motions, since the position of an object involves three space and three orientation coordinates).

To understand the task to be solved, it is useful to distinguish two different conceptual levels. The first level concerns *kinematics*: what arm configuration is needed to bring the end effector to a desired location and into a desired orientation?

For an answer to this question it is useful to introduce the concepts of *task space* and *configuration space*. The task space is our familiar, three dimensional space, augmented by three further coordinates to describe the orientation of an object[2]. It is (usually) the basis for description of what the robot has to do (e.g., move object A along a straight line into position B). The *configuration space*, on the other hand, consists of the possible arm joint settings (e.g., six angles, if the arm has only revolute joints) that determine the arm configuration.

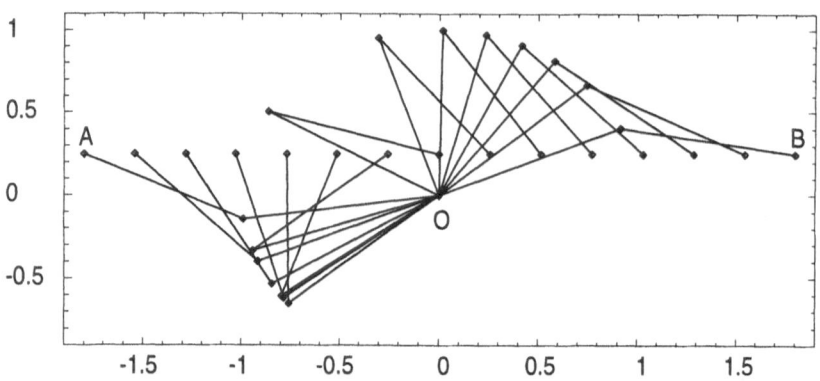

Figure 3: Simulated planar arm with two segments ("shoulder" joint at origin O), traversing with its tip a straight horizontal line from A to B.

The question for the required arm configuration can then be rephrased as that of finding the general mathematical relationship between an arbitrary, given point in task space (six coordinates) and the associated point in configuration space (six joint values; if one goes from configuration to task space, the necessary transformation is called forward kinematics, and inverse kinematics for the opposite direction). Usually, this relationship is highly nonlinear, that is, if we want to move the end-effector along a straight line, the required motion in configuration space may follow a rather complicated, non-linear trajectory (Figs. 3 and 4 show an example for a very simple, planar arm with only two joints; in this case, task and configuration space are both 2-dimensional and can be easily depicted). Moreover, simple looking changes to a trajectory in task space (such as a translation that does not affect the shape of the trajectory) may lead to rather complicated changes of the trajectory in configuration space. Additionally, the transformation between task- and configuration space usually is not one-to-one: for instance, already for the planar arm there may be two different configurations that belong to the same end-effector point in task space. Furthermore, there may be so-called singularities: in these points, a non-zero velocity in configuration space will correspond to a zero velocity in task space, leading to severe complications for controlling arm movements in the vicinity of such points.

The necessary inverse kinematics transformations usually can not be written down as explicit formulas but instead have to be solved by iterative techniques. This has for some time been a source of much discussion, how the necessary computations can be implemented efficiently, and how they might be carried out in the nervous system. At least from a technical perspective, computer processors are now sufficiently fast that we need not worry any longer about this.

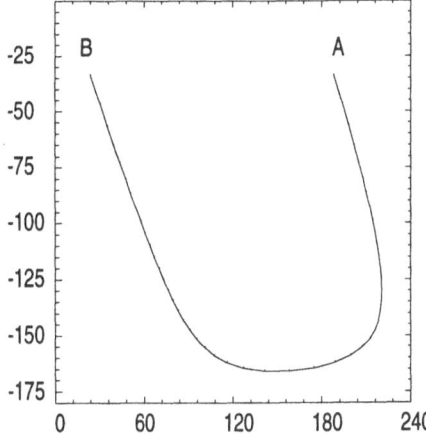

Figure 4: The straight movement in Fig. 3 appears as a strongly curved path in the configuration space spanned by the two joint angles.

A second, and more severe level of difficulties arises from *dynamics*: here we ask, what *forces* and *torques* do we require to accelerate the arm in a desired way (the reader may recall that already a circular movement with constant angular velocity requires maintaining a steady acceleration towards the center of the circle). While kinematics is based on geometrical considerations alone, dynamics requires us to consider additional properties, such as the mass distribution inside the arm, and to connect it via a set of coupled differential equations with the shape and the velocity profile of the desired arm trajectory. Already the differential equations for an arm with only two links are rather complex, and the explicit equations for a 6-degree of freedom PUMA robot arm can only be obtained with the aid of a computer algebra system (they involve more than 50,000 terms).

This has made it for quite some time difficult to understand, how the necessary computations can be carried out in the nervous system so effortlessly, when their mathematical formulation leads to equations of such enormous complexity. It meant a major advance when it was found that the entire computation can be reorganized into a recursive form (Hollerbach 1980), so that the entire complexity can be reduced to a number of operations that only scales quadratically with the number of involved joints (Walker & Orin 1982). To give an idea of this approach, consider evaluation of the expression

$$y = 1 + \frac{x}{2} + \frac{x^2}{4} + \frac{x^3}{8} + \frac{x^4}{16} + \frac{x^5}{32} + \frac{x^6}{64} + \frac{x^7}{128} + \frac{x^8}{256} + \frac{x^9}{512}.$$

From the way the expression is written, it would seem that we would have to evaluate 10 individual terms to find the result y. However, as soon as we discover that the above sum is equivalent to the much more compact expression

$$y = \frac{(x/2)^{10} - 1}{x/2 - 1}$$

the entire computation becomes tremendously simplified. This shows us that the complexity of a task may sometimes be only apparent and due to a poor representation of our computations (of course, in the case of the equations of motion for a robot arm, the economic reorganization of the equations was considerably less obvious than in our illustrative example).

From a more abstract point of view, both, kinematics and dynamics can be viewed as non-linear mappings. While these mappings can become rather non-trivial, robotics research has led to considerable insight into their structure and to the development of efficient computational approaches for their computation. These approaches were in part inspired by biological systems (cf. Chapter 1), and, conversely, may help to better understand aspects of motor intelligence operating at the lowest levels of our motor system.

5. CONTROL

Being able to evaluate the kinematics and the dynamics reasonably fast allows us to make the robot arm follow a desired trajectory if the robot arm were an ideal arm with precisely the properties that were assumed when solving its kinematics and dynamics equations.

In practice, there are many kinds of uncertainties that violate this assumption. Usually, it is difficult to know the precise mass distribution of the moveable parts of a robot arm; it also will change as soon as the robot picks up a heavy object. Furthermore, available actuators can deliver the commanded force or torque only with a certain, limited precision. Friction can occur in many places and poses another source of disturbances that can at best be known approximately; cable stiffness contributes with further changing forces that may be hard to compute analytically.

Robotics is faced with numerous kinds of apparently "dirty" side-effects like those mentioned above, and dealing with them might appear as engaging in an unrewarding activity of just "troubleshooting" in a complex domain.

However, this would be a gross misperception. Actually, robotics has taught us that the successful coping with uncertainties like these is one of the major driving forces for developing a viable basis for intelligent systems. This is very analogous to the situation in nature: it is precisely the high variability and unpredictability of many details in a natural environment that has forced evolution to endow animals with some degree of intelligence (after all, one characterizing feature of intelligence is the ability to deal with the unexpected!). The fact that most of our robots still can do useful work only within a narrow range of carefully controlled situations reflects their still rather low degree of intelligence.

The concept of *feedback control* (Doyle et al. 1992; Jacobs 1993) has been one of the key ideas for dealing with many kinds of uncertainty, and, therefore, is one of the concepts that is of prime relevance to both robotics and prerational intelligence.

The basic idea is simple: any error observed during a movement is converted into a suitable "feedback signal" for the actuators of the robot. The feedback signal must be devised such that it acts in the direction of a reduction of the error. Therefore, one also speaks of "negative feedback" control.[3]

In its simplest form, we encounter the principle of negative feedback in the law of exponential relaxation of a quantity $x(t)$ towards a target value a over time. This law is given simply by

$$x(t) = a + (x(0) - a) \exp(-kt)$$

and one easily verifies that $x(t)$ obeys the simple first-order differential equation

$$\dot{x} = -k \cdot (x - a)$$

which describes exponential decay, which is followed by all linear decay processes, such as, e.g., the cooling of a heated body. The terms of the above equation directly exhibit the concept of negative feedback: $x - a$ measures the deviation ("error") from the target value a, and $-k \cdot (x - a)$ is the feedback signal to cause a corresponding rate of change $\dot{x}$ towards a reduction of the error value. The parameter k is referred to as "feedback gain" and determines the strength of the feedback signal (and, consequently, the speed with which the error is compensated).

Slightly more elaborated variants of the same differential equation (extended to several variables and a matrix of feedback gains k) form the basis of the large majority of conventionally used robot controllers. Important refinements are the inclusion of velocity error signals in addition to positional errors (leading to the concept of PD-controllers, where P denotes the position error proportional term, and D the differential contribution given by the velocity errors), and the further use of an integrated position error to combat slow drifts ("PID-controllers").

A further major issue in control is the question of *stability*. It may happen that it cannot be guaranteed that the feedback signal acts always in the direction of error reduction, but sometimes (e.g., in the case of a fast periodic movement and the presence of delays) it also may contribute towards an amplification of the error. If this happens, the system may become unstable, a situation that must be carefully avoided.

This is a particularly severe problem for biological motor control systems, since the rather long time constants of nervous tissue lead to typical delay times on the order of 50–120 ms (in humans) until feedback signals arrive at their destination. A well-studied example is the stabilization of our vertical body posture. Here, both relatively fast spinal reflexes (about 50 ms latency) and slower reflexes involving various brain centers ("supraspinal reflexes", 120 ms latency) contribute to jointly achieve a stabilization of the otherwise inherently instable body posture (Nashner 1976; Kandel & Schwartz 1985).

Such delays limit allowable feedback gains to rather low values if stability is to be preserved. This has led to extensive debates about how far the concept of negative feedback can provide a basis for the control of human movement. One major proposal has been the equilibrium-point hypothesis: according to this hypothesis, the feedback is not generated explicitly by means of a neural control system. Instead, the stabilizing forces arise from the spring-like properties of stretched muscles, and the nervous signals are only used to prepare suitable equilibrium muscle lengths and tensions such that a desired equilibrium configuration results (Feldman 1966). More recent research has led to several refinements of this proposal (Shadmehr 1995), among them the usage of configuration dependent equilibrium tensions ("force fields" (Shadmehr et al. 1993)) and a mixture of control strategies, relying on the equilibrium point control only to a certain extent (Kawato 1990).

Recent advances in processing power have made it possible also to investigate computer vision as a basis for feedback control ("visual servoing") (Cutzu & Edelman 1994; Hager & Hutchinson 1996). Since image processing is much more time consuming than processing the usually very low-dimensional force- and joint sensor signals, we now also become confronted in the technical domain with considerable feedback latencies, bringing in similar time constraints as for biological motor systems.

Developments in control theory provide potential answers for suitable control strategies. Two related approaches in this regard are inverse model control and model reference control (Narendra & Parthasarathy 1990). To understand these approaches, it is useful to take the view of systems theory, considering the motor system as a "plant" that transforms its input signals

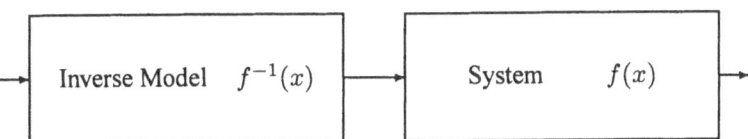

Figure 5: If an inverse model (left box) of the to-be-controlled system (right box) is available, their concatenation behaves like the identity transform and desired output trajectories become very easy to control. If the inverse model is only approximate, the simplification may still be sufficient to good results with a linear controller.

in some (non-linear) way into output signals (the movements or their consequences, as registered by sensors) (Fig. 5). If one can construct the (at least approximate) inverse of that transform, we can use it to map the desired outputs into the required inputs that produce these outputs as consequences, when applied to the motor plant. In other words, for a two-stage system that consists of the inverse model as its first stage, followed by the actual plant as its second stage, the required input becomes very simple: it is just the same as the desired output. Since the inverse model will never be exact, but only approximate, this simple relationship will hold only approximately, but the behavior of the composed system is close to the "identity mapping" and, therefore, close to linear. As a result, a good feedback control scheme is much easier to set up than for the original, often very non-linear, motor plant alone. If the inverse model is fairly accurate, there will be only small errors, so that we may succeed with small feedback gains, thus alleviating the problems that are connected with delays[4].

A closely related approach is *model reference control*. The set-up is analogous, but the inverse model is somewhat modified. This modification is made such that the overall characteristic is no longer (an approximation to) the identity transform, but a prescribed "reference model", for which a feedback controller with certain desirable properties can be realized. The motivation for this is that the construction of a good inverse model may be too demanding for the given plant (we so far did not stress the fact that in our context "inversion" does mean the inversion of a *dynamical system*, which is considerably more difficult than finding the inverse of an ordinary function). However, if we allow the behavior of the composed system to be somewhat more complicated than just the identity transform (but still of a kind that we can well cope with), the task of constructing a suitable transform for the first stage of the system may become significantly facilitated and, thereby, solvable.

The advantages of both approaches result from a *combination* of the diametral concepts of purely feedforward control (no feedback signals are used, everything is computed beforehand: "motor programming") and feedback control (all essential adjustments result from feedback corrections; the feedforward model may be as simple as a purely kinematic model). In the feedback-linearization approach the inverse model implements a feedforward control scheme; remaining errors due to a limited model accuracy are then compensated via feedback control.

A further, third approach also relies on an internal model, but this time it is a *forward* model of the system that is to be controlled. Such a model mimics the behavior of the real system and can be used to predict the effects of motor commands without needing to wait until the feedback signals from the real system become available. Then, the feedback signals from the forward model can be used to realize a feedback controller that is subject to smaller latencies and that, therefore, can work with higher feedback gains, resulting in a better elimination of errors.

By spurring the development of various control architectures, like those sketched above, robotics has also helped to shed light on the question how the brain might control our movements, which essentially swayed between the extremes of a motor program, exemplified by pure feedforward control, and the opposite strategy of a pure feedback-control, e.g. implemented in the form of centrally prescribed muscle tensions or explicit feedback loops from the spinal or supraspinal levels.

6. AUTONOMY

While each of the robot capabilities discussed so-far is sufficiently non-trivial in its operation to make us inclined to use the term "prerational intelligence" to indicate the complexity of the involved processes, a more restrictive usage of this term would be targeted only at processes that qualify as behaviors of the entire robot itself.

One important aspect that then would most likely be associated with pre-rational or even rational intelligence is *autonomy*. By this we mean that the behavior of the robot does not rely on a detailed external specification of a task. Instead, the robot should be able to set its own "goals" or to fulfil a task on the basis of only a coarse specification.

Because of its fundamental importance, one of the tasks that is most frequently studied in the context of autonomous robots is that of *navigation* (cf. Ritter et al. 2000). The ability to traverse an unknown or only partly known environment, relying on visual or other cues to find the way back and to avoid collisions with obstacles, is an indispensable prerequisite for many other activities. Therefore, the ability of navigation can be observed in many species, such as insects, fishes, birds and mammals.

From a computational point of view, an important issue for navigation is the representation of the surrounding space in a way that supports the movement towards a desired goal location. In the simplest cases, it may be sufficient to endow a robot with a small set of (well-chosen) "reactive" behaviors, i.e., behaviors that are very directly triggered by some sensors, without much "deliberation" in between. Examples are the "Braitenberg vehicles" (Braitenberg 1984), little two-wheeled robots with (usually) a pair of light-sensitive sensors at their front. Some simple circuit in the vehicle amplifies the excitation of each sensor and combines both signals in various simple ways into two drive signals for the rear motors. As Braitenberg has discussed in great detail, in the presence of light sources (which may be stationary or even mounted on other Braitenberg vehicles), such simple "agents" can already exhibit remarkably complex behaviors, including simple forms of homing, obstacle avoidance or of seemingly "aggressive" behavior towards other vehicles.

The investigation of these and various similar, simple robots (see, e.g., Maes (1991) for a collection of papers on the subject) has led to the observation that besides explicit programming of the necessary "robot intelligence" to solve a particular task there may be alternative, more indirect ways of implementing many non-trivial capabilities (e.g., obstacle avoidance or homing). These rely on the setting up of a suitable "coupling" between the robot and its environment such that this coupling leads to a dynamics which effects the desired task.

For instance, to approach a goal while at the same time avoiding obstacles can be achieved by just following a force field which is the gradient (the direction of steepest ascent) in an "energy landscape" whose shape has been chosen such that the goal is a very high peak, while obstacles are modelled as "sunken troughs", while the height of any other point in the landscape is taken to be the average of the heights of all the surrounding points in a small (circular) neighborhood region around it. While in the described form this method is only suitable when the shape of the robot can be ignored (e.g., for a sufficiently small robot and large obstacles), the underlying "potential field approach" has been extended to much more complicated situations (Khatib 1991; Khosla & Volpe 1988).

However, the potential field approach already requires the availability of a *map* of the environment, on which obstacle and goal locations can be distinguished. How such maps might be created from raw sensory data, particularly when the data are corrupted by noise, has been the subject of many studies in robotics, and many different algorithms have been devised (see, e.g., Elfes 1987, Thrun 1988, Kurz 1992, and Borenstein et al. 1996 for some representative approaches). The task is also closely linked with the subject of *learning*, another important issue, which we have for lack of space omitted in our discussion so far (but see Braun and Ritter (2000) for a variety of approaches to learning).

With regard to intelligence, one may speculate that the formation of maps is an essential step also for many forms of rational intelligence. In artificial intelligence research, logical deduction and reasoning are often formulated as taking place in suitable, abstract "search spaces", and good representations (maps!) of such search spaces are often crucial for the success of a particular method. So the task of navigation may well be the first prototype of explicit problem solving that has been developed by natural evolution.

Besides maps, another crucial issue for autonomous behavior is a suitable *architecture* by which the proper coordination of more basic activities can be achieved. As already mentioned, for a long time it was taken as granted that such coordination must inevitably be based on an explicit world model in which a planning module would carry out tentative actions, make predictions and used their outcomes as an aid to find a suitable course of action.

Biologically inspired work, aiming at a robust imitation of basic activities, such as navigation and food seeking, has revealed that there are interesting alternatives (Brooks 1991). The perhaps most widely known proposal has been made by Brooks (1986), advocating a layered architecture of behaviors which interact such that the "higher" behaviors can inhibit or modulate the "lower" ones for their purposes. For instance, when the lowest level implements an unbiased wandering activity of the robot, the next higher level might "filter" the movements of the lower level to bias them towards a particular direction when a "goal" comes into view, whereas an even higher level with still higher priority might implement obstacle avoidance over short distances. Due to the nature of the relationship among the different levels, the resulting architecture has been named "subsumption architecture" and has turned out to provide a very good basis for the realization of quite a number of interesting, non-trivial robot behaviors.

While it might seem attractive that approaches such as the subsumption architecture can realize many capabilities in the form of an "emergent behavior" which apparently relieves us of the burden of programming, a closer look rapidly reveals that the design of suitable interaction patterns among the participating modules and the outer world can be a highly non-trivial task. Therefore, subsequent research has attempted to facilitate this task by developing learning strategies to acquire new or refine only coarsely existing skills by exploration (the literature on such approaches has become vast, we cite only a few examples: Zrimec & Mowforth 1991; Verschure et al. 1992; Salganicoff et al. 1996; Walter & Ritter 1996). Other approaches aim at a combination of reactive and deliberative strategies to "take the best of both worlds". Research in this field is currently a largely experimental field, because robot architectures are usually much too complex to allow their in-depth mathematical analysis. Therefore, we most likely will be forced to carry out many small "evolutionary steps" of our artificial robot architectures, not entirely different from the way nature did her job.

7. CONCLUSIONS

Robotics helps us to try out new ideas about intelligence in domains which are simply too complex for a purely theoretical analysis and in which our "rational" expectations may even be misleading. In this way, robotics has already helped us to get a more balanced view of the roles of rational and of prerational intelligence for the shaping of behavior and action in the real world. Before a robot can use any of the symbol and deduction-based methods that have been developed in AI research in an attempt to mimic parts of our rational thinking, it first has to perform a considerable amount of highly non-trivial, "prerational" processing on the raw signals received from its various sensors. To achieve this seems to require methods that appear rather different from the mental activity of reasoning. Most signals are given as continuous-valued quantities, and sometimes, such as in vision, there may be huge numbers of them, ranging in the millions. In this case, each single signal may be highly redundant, given all the others; still we cannot work with a minimal subset, since real world sensor signals are inherently noisy. This has driven the development of methods for signal filtering, noise reduction and integration of many highly redundant, but noisy measurements into a smaller number of more stable and reliable quantities. These then may become suitable for deriving symbols and making contact with the more "classical" methods of AI.

But also here, robotics research has helped to sharpen our awareness for tempting, but problematic assumptions that underlay the whole approach of reasoning, namely the availability of a reliable and consistent world model. It has directed our attention at the important capability of dealing with uncertainty, an ability which now seems to be one of the central features of intelligence within many domains.

One may argue that the many sources of uncertainty with which any animal has to cope in nature may have formed a major driving force, first for the evolution of prerational intelligence, and then, when this had evolved to a level where sufficiently sophisticated preprocessing could provide an "inner world" of more stable and reliable entities, for its subsequent refinement by the various forms of rational intelligence of which we usually are much more aware.

The attempt to extend the use of robots to more and more applications, such as construction sites, forests, the sea floor, or outer space, but also inside ordinary buildings where the environment is primarily adjusted to the needs of humans, such as in hospitals or in households, is likely to exert an analogous evolutionary pressure on the development of artificial robots and we may speculate that this will lead to a similarly staged development of robot intelligence.

We may even directly attempt – to some extent – to simulate evolution. We can build simple robots, subject them to certain tasks and observe their fitness. We can then make changes and favor the fitter machines, investigating whether a similar improvement as in natural evolution occurs over generations, and, particularly, whether and how the "intelligence" in such machines evolves. This approach has in fact been taken, using simulated robots initially, and more recently, real ones (Koza 1992; Ghanea-Hercock & Fraser 1994; Nordin & Banzhaf 1997). Although this line of research must still be considered to be largely in an infancy state, there are already promising results, such as the optimization of foraging behavior.

It may well be that the rapidly expanding internet provides a much better biotope for the evolution of robots than conventional industry or the more direct attempts to mimic evolution. This may first seem paradoxical; after all, there are no "real" robots in the internet. However, there can be all kinds of simulated robots, and they can evolve much faster, since their "material" is software, which can be much more easily copied and modified than any real material we know of.

We already witness the rapid evolution of various "species" of them, with capabilities that are carefully specialized to their particular "niche". Some of them exist for the purpose of populating "virtual realities" with creatures, sometimes with the aim to provide humans with the possibility of a "virtual presence" (termed an "avatar") in a virtual world. Since the human owner of an avatar cannot possibly control all its detailed motions, coming generations of these software robots will have to implement quite a number of those strategies that also underly the motion control of real robots. Similar demands drive the evolution of virtual robots for use as actors in computer generated movies, which is a rapidly increasing market. Another type of virtual robots is being developed as "virtual pets" for the everyday household, and for those who object that a virtual robot is not the real thing at least one large Japanese company has developed a very sophisticated four-legged "robot animal" with stereo vision and highly developed movement capabilities. Finally, we will see the evolution of "information-robots", with senses purely directed at the various information sources in the internet. There are already the various forms of rather simple document search "robots", but the inclusion of materials such as digital images for search requests will require endowing these software robots with vision capabilities much like our own, although an internet search robot may never have the need to be connected to a camera to view the "world outside". It will be interesting to see how the advent of such strange seeming "robot species" and their capabilities will shape our judgement of intelligence, both in the rational and in the prerational domain.

Universität Bielefeld, Germany

NOTES

[1] Of course, the activation of each artificial neuron follows a precisely specified and known mathematical rule; however, the level at which this rule is formulated is very distant from a level at which one can speak of any meaningful object features. Therefore, it makes little sense to consider relations and tranformations of such features as encoded in the neural transformation rules.

[2] In robotics, the resulting six coordinates are usually combined into a 4×4 matrix of "homogeneous coordinates", which makes many operations mathematically more convenient. For lack of space, we will not consider this technical point any further, the interested reader is referred to any standard textbook on robotics, e.g., Paul (1981), Fu et al. (1987).

[3] This is to be carefully distinguished from the much more recent concept of "positive feedback" advocated by Cruse (Kindermann et al. 1995) as a means for achieving certain forms of intelligent sensori-motor coordination.

[4] There is another approach to ensure stability even for highly non-linear and uncertain systems, named "sliding mode control". Space does not permit consideration of the underlying ideas, but the reader may wish to consult Utkin (1992).

REFERENCES

Agre, P.E., & D. Chapman (1990). What are plans for? In P. Maes (ed.), *Designing autonomous agents* (pp. 17–34). Cambridge, MA: Elsevier/MIT Press.

Alligood, K.T., T.D. Sauer, & J.A. Yorke (1997). *Chaos. An introduction to dynamical systems*. New York: Springer.

Arrowsmith, D.K. (1994). *An introduction to dynamical systems*. Cambridge, UK: Cambridge University Press.

Ballard, D. (1991). Animate vision. *Artificial Intelligence* **48**, 57–86.

Borenstein, J., Everett, B., & Feng, L. (1996). *Navigating mobile robots: Systems and techniques*. Wellesly, MA: A.K. Peters Ltd.

Brady, M. (1989). *Robotics science*. Cambridge, MA: MIT Press.

Brady, M., J. Hollerbach, T. Johnson, T. Lozano-Pérez, & M. Mason (eds.), (1982). *Robot motion: Path planning and control*. Cambridge, MA: MIT Press.

Braitenberg, V. (1984). *Vehicles: Experiments in synthetic psychology*. Cambridge, MA: MIT Press.

Braun, H., & H. Ritter (2000). Introduction to Part XI: Plasticity and learning. In H. Ritter, H. Cruse, & J. Dean (eds.), *Prerational intelligence: Adaptive behavior and intelligent systems without symbols and logic*, Vol. 2 (pp. 589–594). Dordrecht, The Netherlands: Kluwer Academic Publishers.

Brooks, R.A. (1986). A robust layered control system for a mobile robot. *IEEE Journal of Robotics and Automation* **2**(1), 14–23.

Brooks, R.A. (1989). The whole iguana. In M. Brady (ed.), *Robotics science*. Cambridge, MA: MIT Press.

Brooks, R.A. (1990). Elephants don't play chess. In P. Maes (ed.), *Designing autonomous agents* (pp. 3–16). Cambridge, MA: Elsevier/MIT Press.

Brooks, R.A. (1991). Intelligence without representation. *Artificial intelligence* **47**, 139–159.

Bülthoff, H., & S. Edelman (1992). Psychophysical support for a 2d view interpolation theory of object recognition. *Proceedings of the National Academy of Sciences, USA,* **89**, 60–64.

Cruse, H. (1996). *Neural networks as cybernetic systems*. Stuttgart, Germany: Thieme Verlag.

Cutzu, F., & S. Edelman (1994). Canonical views in object representation and recognition. *Vision Research* **34**, 3037–3056.

Dario, P., P. Ferrante, G. Giacalone, L. Livaldi, B. Allotta, G. Buttazzo, & A.M. Sabatini (1992). Planning and executing tactile exploratory procedures. *IEEE International Conference on Intelligent Robots and Systems* (pp. 1896–1903). Raleigh, North Carolina.

Doyle, J.C., B.A. Francis, & A.J. Tannenbaum (1992). *Feedback control theory*. Prentice-Hall, NJ: Macmillan.

Edelman, S., D. Reisfeld, & Y. Yeshurun (1992). Learning to recognize faces from examples. Proceedings of the 2nd European Conference on Computer Vision (S. Margherita Ligure, Italy). *Springer Lecture Notes in Computer Science* **588**, 787–791.

Elfes, A. (1987). Sonar-based real-world mapping and navigation. *IEEE Robotics and Automation Magazine* **3**, 249–265.

Feldman, A.G. (1991). Functional tuning of the nervous system with control of movement or maintenance of a steady posture. II: Controllable parameters of the muscles. *Biophysics* **11**, 565–578.

Fikes, R.E., & N.J. Nielsson (1971). Strips: A new approach to the application of theorem proving to problem solving. *Artificial Intelligence* **2**, 189–208.

Flash, T., & N. Hogan (1985). The coordination of arm movements: An experimentally confirmed mathematical method. *Journal of Neuroscience* **5**(7), 1688–1703.

Fu, K.S., R.C. Gonzalez, & C.S.G. Lee (1987). *Robotics: Control, sensing, vision, and intelligence*. New York: McGraw-Hill.

Ghanea-Hercock, R., & A.P. Fraser (1994). Evolution of autonomous robot control architectures. In T.C. Fogarty (ed.), *Lecture Notes in Computer Science* **865**: *Evolutionary Computing*. Heidelberg: Springer.

Hager, G., & M. Mintz (1991). Computational methods for task-directed sensor data fusion and sensor planning. *International Journal of Robotics Research* 10, 285–313.

Hager, G., & Hutchinson, S. (eds.), (1996). *IEEE Robotics and Automation.* Special issue on visual servoing.

Heidemann, G., & H. Ritter (1996). A neural 3d-object recognition architecture using optimized gabor filters. *Proceedings of the 13th International Conference on Pattern Recognition* (pp. 70–74). Wien: IEEE Computer Society Press, Los Alamitos, CA.

Hollerbach, J.M. (1980). A recursive lagrangian formulation of manipulator dynamics and a comparative study of dynamics formulation complexity. *IEEE Transactions on Systems, Man and Cybernetics* **SMC-10**, 730–736.

Horn, B.K. (1977). Understanding image intensities. *Artificial Intelligence* **8**, 201–231.

Hurlbert, A., & T. Poggio (1988). Making machines (and artificial intelligence) see. In S.R. Graubard (ed.), *The artificial intelligence debate – False starts, real foundations* (pp. 213–240). Cambridge, MA: MIT Press.

Jacobs, O.L.R. (1993). *Introduction to control theory.* Oxford, UK: Oxford Science Publications.

Johansson, R.S. (1996). Sensory control of dextrous manipulation in humans. In A.M. Wing, P. Haggard, & J.R. Flanagan (eds.), *Hand and brain* (pp. 381–414) New York:. Academic Press.

Kandel, E., & J. Schwartz (1985). *Principles of neural science.* New York: Elsevier.

Kawato, M. (1990). Computational schemes and neural network models for formation and control of multijoint arm trajectory. In W.T. Miller, R.S. Sutton, & P.J. Werbos (eds.), *Neural networks for control* (pp. 197–228). Cambridge, MA: MIT Press.

Khatib, O. (1991). Real-time obstacle avoidance for manipulators and mobile robots. In S.S. Lyengar & A. Elfes (eds.), *Autonomous mobile robots: Perception, mapping and navigation* (Vol. 1) (pp.428–436). Los Alamitos, CA: IEEE Computer Society Press.

Khosla, P., & R. Volpe (1988). Superquadric artifical potentials for obstacle avoidance and approach. *Proceedings of the 1988 IEEE International Conference on Robotics and Automation* (pp. 1778–1784). Philadelphia, PA.

Kindermann, T., H. Cruse, & Ch. Bartling (1995). High-pass filtered positive feedback: decentralized control of cooperation. In P. Chacon, F. Moran, A. Moreno, & J.J. Merelo (eds.), *Advances in artificial life* (pp. 668–678). Heidelberg: Springer.

Koza, J.R. (1992). Evolution of subsumption using genetic programming. In F.J. Varela & P. Bourgine (eds.), *Proceedings of the 1st European Conference on Artificial Life* (pp. 110–119). Cambridge, MA: MIT Press.

Kurz, A. (1992). Building maps for path planning and navigation using learning classification of external sensor data. In I. Alexander & J. Taylor (ed.), *Artificial neural networks 2* (pp. 587–590). Amsterdam: Elsevier.

Littmann, E., A. Drees, & H. Ritter (1996). Visual gesture-based robot guidance with a modular neural system. In D. Touretzky, M. Mozer, & M. Hasselmo (eds.), *Advances in neural information processing systems 8* (pp. 903–909). Cambridge, MA: MIT Press.

Logothetis, N.K., J. Pauls, H.H. Bülthoff, & T. Poggio (1994). View-dependent object recognition by monkeys. *Current Opinion in Biology* 4, 401–414.

Lozano-Perez, T. (1987). A simple motion-planning algorithm for general robot manipulators. *IEEE Journal of Robotics and Automation* **RA-3**(3), 224–238.

Lozano-Perez, T., J.L. Jones, E. Mazer, & P.A. O'Donnell (1989). Task-level planning of pick-and-place robot motions. *Computer* **22**(3), 21–29.

Maes, P. (ed.), (1991). *Designing autonomous agents.* Cambridge, MA: Elsevier/MIT Press.

Marr, D. (1982). *Vision.* San Francisco, CA: Freeman.

Massone, L., & E. Bizzi (1989). A neural network model for limb trajectory formation. *Biological Cybernetics* **61**, 417–425.

Murase, H., & S. Nayar (1995). Visual learning and recognition of 3d-objects from appearance. *International Journal of Computer Vision* **14**, 5–24.

Narendra, K.S., & K. Parthasarathy (1990). Identification and control of dynamical systems using neural networks. *IEEE Transactions on Neural Networks* **1**, 4–27.

Nashner, L.M. (1976). Adapting reflexes controlling the human posture. *Journal for Experimental Brain Research* **26**, 59–72.

Nayar, S.K., & T. Poggio (1996). *Early visual learning.* Oxford, UK: Oxford University Press.

Nölker, C., & H. Ritter (1998). Illumination independent recognition of deictic arm postures. *Proceedings of the 24th Annual Conference of the IEEE Industrial Electronics Society* (pp. 2006–2011). Aachen, Germany: IEEE Computer Society Press, Los Alamitos, CA.

Nordin, P., & W. Banzhaf (1997). An on-line method to evolve behavior and to control a miniature robot in real time with genetic programming. *Adaptive Behavior* **5**, 107–140.

Paul, R. (1981). *Robot manipulators: Mathematics, programming, and control.* Cambridge, MA: MIT Press.

Poggio, T., V. Torre, & C. Koch (1985). Computational vision and regularization theroy. *Nature* **317**(6035), 314–319.

Richards, W. (ed.), (1988). *Natural computation.* Cambridge, MA: MIT Press.

Ritter, H. Cruse, & J. Dean (eds.), (2000). Introduction to Part IX: Navigation. In H. Ritter, H. Cruse, & J. Dean (eds.), *Prerational intelligence: Adaptive behavior and intelligent systems without symbols and logic*, Vol. 2 (pp. 361–365). Dordrecht, The Netherlands: Kluwer Academic Publishers.

Roberts, L.G. (1965). Machine perception of three-dimensional solids. In J.P. Tippet (ed.), *Optical and electro-optical information processing* (pp. 159–197). Reprinted in J.K. Aggarwal, R.D. Duda, & A. Rosenfeld (eds.) (1977), *Computer methods in image analysis* (pp. 285–323). Cambridge, MA: MIT Press.

Salganicoff, M., M. Rucci, & R. Bajcsy (1996). Unsupervised visuo-tactile learning for control of manipulation. In S.K. Nayar & T. Poggio (eds.), *Early visual learning* (pp. 329–362). Oxford, UK: Oxford University Press.

Schuster, H.G. (1988). *Deterministic chaos*. Weinheim, Germany: Verlagsgesellschaft.

Shadmehr, R. (1995). Equilibrium point hypothesis. In M. Arbib (ed.), *The handbook of brain theory and neural networks* (pp. 370–372). Cambridge, MA: MIT Press.

Shadmehr, R., F.A. Mussa-Ivaldi, & E. Bizzi (1993). Postural force fields of the human arm and their role in generating multi-joint arm movements. *Journal of Neuroscience* **13**, 45–63.

Thrun, S. (1998). Learning metric-topological maps for indoor mobile robot navigation. *Artificial Intelligence* **99**, 21–71.

Utkin, V. (1992). *Sliding modes in control optimization*. Berlin: Springer.

Verschure, P.F.J.M., B. Kröse, & R. Pfeifer (1992). Distributed adaptive control: The self-organization of structured behavior. *Robotics and Autonomous Systems* **9**, 247–265.

Walker, M.W., & D.E. Orin (1982). Efficient dynamic computer simulation of robotic mechanisms. *Journal of Dynamic Systems, Measurement and Control* **104**, 205–211.

Walter, J., & H. Ritter (1996). Rapid learning with parametrized self-organizing maps. *Neurocomputing* **12**, 131–153.

Waltz, D.I. (1972). *Generating semantic descriptions from drawings of scenes with shadows*. PhD thesis, AI Lab. Cambridge, MA: MIT Press.

Weng, J.J., N. Ahuja, & T.S. Huang (1993). Learning recognition and segmentation of 3d objects from 2d images. *Proceedings of the International Conference on Computer Vision* (pp. 121–128). Berlin: IEEE Computer Society Press, Los Alamitos, CA.

HEINRICH BRAUN

COMPUTER SCIENCE PERSPECTIVES ON PRERATIONAL INTELLIGENCE

1. INTRODUCTION

The subject of computer science is information processing or problem solving with electronic computing devices. These devices may be standard personal computers, large parallel computers or special purpose hardware circuits.

Some topics already thoroughly discussed in *The Mathematical Perspectives on Prerational Intelligence* (Ritter & Bauer, this volume) are important and central to the concept of prerational intelligence from the perspective of computer science. These topics include the following:

- *Coding*: The representation of information in neural networks.
- *Learning*: The construction and tuning of neural networks.
- *Optimization*: A fundamental task for problem solving.
- *Pattern recognition*: In a fundamental sense, the most important application of neural networks.

Therefore, I want to focus my considerations on the following four aspects:

1. *Models for describing algorithms of prerational intelligence*: I will argue that threshold circuits are the adequate model for formalizing prerational information processing in the brain and that they are equivalent to the standard artificial neural networks, whereas the Turing machine or the von Neumann architecture is the adequate model for formalizing algorithms on the standard computer (PC or workstation). In Section 2, I discuss models for computability and conclude that this model is too general to distinguish between rational and prerational intelligence. Therefore, I discuss in Section 3 more sophisticated models which take aspects of efficiency and parallelization into consideration.

2. *Problem Solving*: What are the fundamental paradigms of computer science for problem solving and to what extent are they used for realizing prerational intelligence? In Section 4, I will discuss a variety of heuristics, including hillclimbing, gradient descent, evolution, divide-and-conquer, branch-and-bound, recursion and dynamic programming.

3. *Synthesis of neural networks*: What are the fundamental methods of computer science for constructing programs and to what extent are they used

J. Dean et al. (eds.), *Prerational Intelligence: Interdisciplinary Perspectives on the Behavior of Natural and Artificial Systems*, 161–200.

for realizing prerational intelligence; what methods are special for constructing neural networks? In Section 5, I will discuss the concepts of modularization, information hiding, hierarchization of information processing and divide-and-conquer as standard methods in computer science whereas learning and evolution are non-standard methods commonly used for constructing neural networks.

4. *Strategy Learning*: How can we learn to optimize our behavior in a given environment? In Section 6, I will discuss three scenarios: The first scenario is supervised learning, where the learner tries to imitate the strategy of an expert. The expert gives prototypical examples from which the learner has to generalize a complete strategy. The second scenario is goal directed learning where a teacher sets the goals without telling how these can be achieved. The last and most general one is reinforcement learning where the learner tries to optimize a reinforcement signal received from the environment, as, for example, in the case of a foraging honeybee which gets positive reinforcement signals when it collects nectar from a flower and negative reinforcement signals as it loses energy during flight.

2. MODELS FOR COMPUTABILITY

In order to get some insight into natural intelligence and especially into prerational intelligence, we have to make models, i.e., we have to find the essential parameters and to neglect the peculiarities in order to reduce the complexity of the problem.

The highest level of abstraction neglects completely the inner structure of the neurocomputing engine represented by the brain. Finite state machines model the brain as a reflex machine (Mealey 1955) or as a blackbox with an input, an output and an internal state (Moore 1956). Its behavior can be described as follows: in every time step it reads from the input, writes to the output and – depending on its current state and the input read – changes its state. Obviously, this general model can describe the behavior of any computer (and, if one believes in reductionism, even the human brain), because every computer works in time steps, reads from its input device, writes to its output device and possesses a finite memory (i.e., a finite number of states). Nevertheless, this model is very inadequate because a processor with 10 Mbytes memory has more than $2^{10\ 000\ 000}$ different states. In order to describe the behavior of a computer we have to model its logical structure (memory, registers, etc.), but we may neglect the physical structure. The appropriate descriptive language for the logical structure of a computer is well-known (formal languages: context-free grammars for specifying the syntax, context-sensitive (attributed) grammars for specifying the semantics, see any textbook on compiler theory), but finding a corresponding language for the human brain remains the dominant unsolved question. The semantic models of conventional computer programs are completely specified, whereas the meaning of linguistic terms in human language is fuzzy and imprecise. Nevertheless, finite state machines may be appropriate models for small neural or electrical circuits. For these simple machines we have a beautiful theory with constructive solutions for many important questions, such as the following:

- What is the minimal finite state machine (fewest states) for a given behavior?
- Is the behavior of two given finite state machines equal?
- Is a given function computable by a finite state machine?

By ignoring time and space (memory) constraints, Turing, Gödel, Church, Kleene et al. developed the theory of computability, which addresses the question of what kinds of functions are effectively computable by any automaton (e.g., artificial intelligent system). They used three different modeling approaches: the algorithmic perspective (Turing), the mathematical perspective (Kleene) and the logical perspective (Gödel). All three different approaches are equivalent because they define the very same set of effectively computable functions. This fact led Church to his famous thesis, namely, that this set includes exactly all functions effectively computable by any automaton (Church 1941). In the following, we will discuss these three models in more detail.

2.1 *Turing Machine*

Turing (1936) proposed as an algorithmic model his A-machine, now commonly referred to as a Turing machine, a device which can only read from, or write symbols to and move along an infinitely extensible tape (its memory) according to its program, which is just a finite decision table. This machine was designed as a minimalistic approach because its memory is reduced to a linear tape with only incremental access (i.e., no direct address handling) and the alphabet may be limited to just two symbols (bitwise read and write). Nevertheless, this simple machine can compute any effectively computable function, i.e., this model is as strong as any other descriptive, algorithmic language such as PASCAL, C, or ADA, although it lacks all of their convenient facilities for algorithmic specification.

Obviously, each program in any programming language (e.g., PASCAL, C, ADA, or simple Turing tables) can be encoded as a binary sequence of 0's and 1's. This sequence can be interpreted as a binary number. Therefore, for any programming language we can formalize an *interpreter* as an effectively computable function of two operands, the program p (coded as a binary number) and a value x, which computes $p(x)$. Because these interpreters are effectively computable, they, too, can be computed by a Turing machine. Such a Turing machine is called universal. In other words, a universal Turing machine can execute a universal program u which computes not just one special function but *all* effectively computable functions by instantiating the program encoded in the first operand using the second operand as input, that is, $f(x) = u(p, x)$.

This property is the corner stone of the success of computer technology: We can build one universal machine (the hardware) and use it for all different purposes just by installing different programs (the software). Only by harnessing the economic power of millions of users could such universal processors be developed to the impressive performance levels of today's technology. On the other hand, a special purpose machine can be faster than such a universal interpreter. Therefore, most real computer systems contain both universal processors and special purpose circuits. Between the two extremes is a wide range of intermediate information processing machines. The degree of specialization depends mainly on economic reasons: How many users will pay what development cost to achieve increments in speed through specialization.

This comparison leads to the interesting question of the extent to which components of the human brain are universal or special purpose structures. It seems plausible to suggest that the degree of specialization in the brain varies with associated costs in an analogous manner. Moreover, I would suggest that rational thinking is a more or less universal process (just manipulating symbols) whereas prerational intelligence depends on specially tailored processing circuits. In other words, the *rational* machine is the universal processor with which we can compute any function with cheap hardware costs (because it is universal) but possibly with a high cost in time and energy. With this mechanism, an engineer or the self-transforming brain can test new strategies (rapid prototyping) before spending high hardware costs for constructing a special-purpose circuit, i.e., fixing the strategy in neural hardware as prerational intelligence (see Section 6.4)

2.2 *Recursive Functions*

In 1888, Dedekind defined mathematically a class of effectively computable functions which are now called primitive recursive. This approach also may be called minimalistic because it uses as basic functions only the most elementary functions, namely, the constant 0, the successor function $S(x) = x + 1$ and the projection functions $P_i(x_1, .., x_n) = x_i$. Moreover, it uses only two higher operators: composition and primitive recursion:

- A function composed by primitive recursive functions is also primitive recursive.
- A primitive recursive function may be recursively specified if the argument is decremented in this equation as, for example, in the specification of the Fibonacci numbers by the equation $f(x + 1) = f(x) + f(x - 1)$.

More complex functions, like *ADDITION, MULTIPLY, EXPONENTIAL,* etc., can easily be specified by this recursion operator. For example, $ADD(x, y + 1) = S(ADD(x, y))$ and $ADD(x, 0) = x$.

This class already contains every algorithm which can be computed on a computer with an exponential upper space bound (i.e., its memory size can be bounded by an exponential function depending on the length of the input); it contains every function computable in realistic time and space.

Nevertheless, by specifying a function which differs from any primitive recursive function it can be easily shown that there are effectively computable functions which are not primitive recursive. The proof uses the fact that we can effectively enumerate all primitive recursive functions. By this enumeration we can construct an algorithm which computes a function different from all primitive recursive functions, more precisely, a function which differs from the kth primitive recursive function on input k for every k.

In 1936, Kleene defined the μ-recursive functions by substituting μ-recursion for primitive recursion. In terms of modern programming languages, we may call the primitive recursion a "for-loop" (e.g., for $i := 1$ *to n do...*) and the μ-recursion a "while-loop" (e.g., while $i > 0$ *do...*). The class of μ-recursive functions is equal to the class of functions which can be computed by a Turing machine. Moreover, this class defines the set of all functions which can be computed on any computer with unbounded memory because there is at least in principle an interpreter (simulator) on the Turing machine for ADA, C, PASCAL, FORTRAN, or any programming language and vice versa. This fact is the basis of the famous thesis by Church (1936) which states that the set of effectively computable functions is equal to the set of μ-recursive functions, i.e., whatever new programming language we may invent, it will not allow us to formulate an algorithm which cannot already be computed on a Turing machine. Nevertheless, new programming languages may be much more convenient than the minimalistic Turing machine.

If we believe that the human brain is nothing more than a computing engine, we may conclude that every human behavior belongs to the set of

μ-recursive functions. Conversely, every μ-recursive function can be computed by a human brain if we abstract from the individual constraints on time and memory by using several life spans and additional memory capacity such as paper, if needed. Unfortunately, there is one drawback: μ-recursive functions may be only partially defined, which means that the corresponding Turing machine does not terminate on every input. Moreover, it cannot even be effectively decided whether a Turing machine terminates on a given input. It is, therefore, remarkable that the primitive recursive functions do not have such a drawback: they are always total (i.e., overall defined), they have convenient programming languages (just take a language like Pascal, FORTRAN, ALGOL, or C, and exclude the operators "while", "repeat", "go to", and "recursion") and they contain every function computable in realistic time. Nevertheless the μ-recursive functions became the computing paradigm and they are the basis for the computer metaphor as it is commonly applied to the human brain.

2.3 *First Order Predicate Logic*

At the end of the last century, Frege developed a logical calculus for first order predicate logic, in which he introduced in addition to the propositional logic the existence quantor ($\exists$: there exists an instantiation of variable x) and the all quantor ($\forall$: for all instantiations of variable x). These additions allowed many advances. For example, Frege could formalize the law of commutativity for the addition operator using the following formula: $\forall x, y :$ $x + y = y + x$, or, equivalently, $\neg \exists x, y : \neg [x + y = y + x]$ (where $\neg$ negation). Another example is the existence of more than one element denoted by the following formula: $\exists x, y : \neg [x = y]$, or, equivalently, $\neg \forall x, y : x = y$.

The aim of a logical calculus is to formalize deduction or reasoning. Given a set of rules (or an axiom system), we want to formalize the deduction of all valid theorems (statements). A theorem may be correctly deduced (another term is *derived*) from a given rule set if it is valid for all models which obey the rule set. This rule set may be infinite, but in the following we assume that the rule set is effectively constructible, i.e., it is effectively (μ-recursive) decidable whether a formula is an axiom.

Robinson proposed a logical calculus (for first order predicate logic) with only six simple rules (axioms), including as operators only the constant 0 and the successor $S(x)$ ($= x + 1$) such that there exists for every μ-recursive function f a corresponding formula F such that for all x, y the following relation holds:

$$f(x) = y$$

$\Longleftrightarrow F(x, y) \land \forall z [F(x, z) \Longrightarrow y = z]$ is true in the model of natural
numbers,

$\Longleftrightarrow F(x, y) \land \forall z [F(x, z) \Longrightarrow y = z]$ is derivable in the calculus.

In other words, the formula F is a logical formalization of the function f, and, generally, every μ-recursive function is representable in this calculus.

On the other hand, the set of derivable formulas T is effectively enumerable for every (first order predicate) logical calculus with an effectively constructible rule set (i.e., decidable axiom system). To be more specific, there is a μ-recursive function with range $int(T)$, where $int(T)$ is the set of natural numbers we get by interpreting the binary text string coding of each derivable formula as a binary number.

Therefore, we may conclude that all μ-recursive functions f may be effectively computed for each given input x just by enumerating all derivable formulas until we enumerate a formula $F(x, y) \land \forall z [F(x, z) \Longrightarrow y = z]$ for a certain y. The correct output y could then be easily extracted from that formula. Because all deductions in such a logical calculus can also be computed by a human brain and, if we believe in reductionism, vice versa: All human deductions could in principle be formalized by any calculus at least as strong the Robinson arithmetic (which is already strong enough to formalize all μ-recursive functions, see above).

Of course there are more efficient ways to compute functions by a logical calculus than simple enumeration. The programming language Prolog, for example, is based on this paradigm of a deduction engine, but it uses an efficient search strategy to derive formulas (theorems).

Unfortunately, there exists no logical calculus (in first order predicate logic) with an effectively constructible rule set which is complete for the model of natural numbers in the sense that all formulas valid for the model of natural numbers can be derived by this calculus. In his famous incompleteness theorem, Gödel proved that we could construct for every effectively constructible rule set M a formula F_M which is not derivable from M. This formula F_M (used by Gödel in his proof) is just the formalization of the self-reflexive predicate "I am not derivable from M".

This fact led some researchers to the comfortable conclusion that our human brain is more intelligent than every deduction engine (alias Turing machine) because we can construct for any given calculus (at least as strong as the Robinson arithmetic) a formula for which we can prove that it is true for the standard interpretation (model of natural numbers) but which is not derivable in this given calculus. In other words, the human brain is stronger than any logical calculus. Unfortunately, this argument has a caveat: This construction can also be formalized by any logical calculus at least as strong as the Robinson arithmetic. This seems to be a paradox, but the reason lies in the fact that the human brain can only prove that the constructed formula is not derivable under the assumption that the given rule set is valid for the model of natural numbers. But in general the truth of this assumption is not effectively decidable by a Turing machine and it seems plausible that this is also not decidable by the human brain because the rule set (axioms) may include any open mathematical problem as an axiom.

Therefore, the thesis that the human brain is in principle stronger than a deduction engine based on Robinson arithmetic remains to be proven and the thesis of reductionism is not contradicted.

Summarizing, we proposed three different, but equivalent paradigms for the human brain:

- the *computer science perspective*: The brain as a computer with a processor and memory (i.e., the computer metaphor, see operational programming languages like ALGOL, Pascal, C);
- the *logical perspective*: the brain as a deduction engine based on a rule system (see logical programming languages like Prolog);
- the *mathematical perspective*: the brain as a blackbox computing μ-recursive functions (see functional programming languages like Lisp).

All three models neglect the efficiency of computing. Therefore, these models abstract from time and space constraints. Accordingly, parallelization or uncertain (fuzzy) reasoning are of no interest on this level of abstraction. Nevertheless, these models set the limits to more specific models: Any more constrained and more specific model can only compute a subset of the μ-recursive functions.

It seems obvious that on the level of these models (computer metaphor alias deduction engine alias μ-recursive functions) we cannot distinguish between rational and prerational intelligence. So let us look at more specific models.

3. MODELS OF ALGORITHMIC COMPLEXITY

The Turing machine is the minimalistic model of conventional computer systems with what is referred to as a von Neumann architecture: a potentially infinite memory and, as the only active component, a central processing unit (CPU). The input and output of such a machine are strings over a given alphabet. At least since the era of the CD-ROM, everybody knows that we can encode efficiently any information (picture, sound, text) in a binary string. Therefore, we may assume without loss of generality that the input is a binary string (a sequence of $0's$ and $1's$). A von Neumann computer works sequentially because the only active part is the CPU which works step by step looking at the memory through a small but movable window.

If we want to measure the complexity of an algorithm for such a machine, then we want to give upper limits for the computing time and for the memory (space) required. Usually, these limits are measured relative to the input length. If an algorithm has to read (sequentially) the complete input string before it can produce the output, then the required time just for reading will increase linearly with the length of the input. Therefore, the fastest algorithms are "linear in time" and form the linear time complexity class. One example of this class is the algorithm for a parity check, i.e., the output is 1 if and only if the number of 1's in the input is odd.

The classes of algorithms with time complexity bounded by a polynomial of degree k form a hierarchy in k. The union (or supremum) of this hierarchy is the class P containing all algorithms with polynomial time complexity. This class contains more or less all efficiently computable algorithms. Another class is the set of algorithms with exponential time complexity (EXP). The memory (spatial) complexity measure leads to a similar hierarchy, that is, to a class of algorithms (PSPACE) with polynomially bounded complexity in space. An easy proof shows that $P \subset \text{PSPACE} \subset \text{EXP}$.

All these complexity classes are appropriate for measuring the complexity of algorithms tailored for computing models with the von Neumann architecture (von Neumann 1951), but not for computing models with parallel processing and distributed memory. For example, the parity check can be done in logarithmic time by a simple circuit with Boolean gates (processing nodes of the circuit) with two inputs and one output each. Therefore we may state:

- The computer metaphor, the paradigm of a deduction engine and the functional approach which model the human brain as a Turing machine, a logical deduction engine and a μ-recursive function, respectively, are equivalent and appropriate models if we neglect time or space complexity.
- Otherwise, all three models are inadequate.

3.1   *Circuit Complexity*

In order to investigate the time and space complexity of neural circuits, we have to model the processing structure more closely: the brain as a neural network with nodes called neurons or gates. Typically, a biological neuron has many inputs (dendrites), but only one output (axon). From the perspective of computer science it seems to be appropriate to model the input and the output in a binary manner:

- the input from the receptors may be 0 or 1
- the output of a neuron may be 1 if it is active, or 0 otherwise.

Moreover, we may assume that the neuron is deterministic and has no memory, that is, its output is determined by its current input. In this case, each neuron just computes a Boolean function.

McCulloch and Pitts (1943) used such neurons to formalize the first neural network (or neural circuit). The so-called McCulloch-Pitts neuron possesses excitatory and inhibitory inputs and a threshold θ (where θ is a positive integer). It outputs 1 if all inhibitory inputs are 0 and the number of excitatory inputs which are 1 exceeds the threshold θ.

In the following, we discuss the complexity of neural circuits without loops. Obviously, it holds that the whole circuit computes a Boolean function if each of its gates (neurons) computes a Boolean function. The theory of circuit complexity investigates which classes of Boolean functions can be computed by circuits with gates of a given type and with limited network size. Typical gates of classical hardware circuits are AND (output 1 if all inputs are 1), OR (output 1 if at least one input is 1) and NOT (output 1 if the only input is 0). Obviously, we can compute any Boolean function f of input dimension n in only three parallel steps by transforming the set $A = \{x | f(x) = 1\}$ into a circuit with n NOT gates to represent the negated inputs, a hidden layer of at most 2^n AND gates (one for each $x \in A$) and an OR gate as output neuron. For example, if $A = \{110, 011\}$ we transform in three steps:

$$
\begin{aligned}
A \quad &= \quad \{110, 011\} \\
&\rightarrow \quad \text{OR}(110, 011) \\
&\rightarrow \quad \text{OR}(\text{AND}(110), \text{AND}(011)) \\
&\rightarrow \quad \text{OR}(\text{AND}(x_1, x_2, \text{NOT}(x_3)), \text{AND}(\text{NOT}(x_1), x_2, x_3)).
\end{aligned}
$$

The caveat of this construction is the exponential growth of the hidden layer (AND gates) when scaling up the input dimension. If we have just 100 receptors (input neurons) we may need up to 2^{100} AND gates, i.e., more than the number of atoms in the universe. Therefore, we limit in the following the number of gates (neurons) by a polynomial.

The time complexity of a circuit is given by its depth (i.e., the longest path in the circuit). For modeling prerational intelligence it is reasonable to limit the reaction time to about a second. The argument, based on various measured reaction times, is that a longer time would allow rational processes to contribute, whereas an immediate reaction excludes rational considerations. If we further note that a biological neuron typically requires about

10 ms of processing time, we may limit the depth of a circuit to about 100 (in other words: We suppose in the following the well-known 100 step rule).

Typically, the number of receptors far exceeds 100 (see optical receptors in the eye, acoustical receptors in the ear, olfactory receptors in the nose, tactile sensors in the skin). Therefore, linear (reaction) time complexity is too slow for prerational intelligence. Two variants seems to be appropriate: Constant time complexity and logarithmic time complexity, because both fulfill the 100 step rule (logarithmic time complexity up to 2^{100} receptors which is more than sufficient for biological brains). In the following, I summarize the important results for both variants. For further readings I recommend the book by Ian Parberry (1994).

3.2 Neural Circuits With Constant Depth

The aim of circuit complexity theory is to give upper bounds for the network size (space complexity) and depth (time complexity). The number of inputs (*fan-in*) of each gate must be potentially unbounded. Otherwise it would not be possible to utilize the whole input of the circuit in constant depth such as depth 2, for instance. Examples of such neurons (or gates) have already been mentioned: AND and OR with arbitrary *fan-in* (reminder: NOT has only *fan-in* 1). Other examples are MAJORITY (output 1 if the majority of the inputs is 1) and THRESHOLD (output 1 if the number of inputs with 1 is greater than the threshold).

For each given type of neurons we can define the class of Boolean functions which are computable in depth k and polynomial size. Obviously, these classes form hierarchies within each type (i.e., the class of depth k is a subset of the class of depth $k + 1$). It is an interesting question whether this hierarchy is proper, meaning that the class of depth k is a proper subset of the class of depth $k + 1$ for all k, or whether this hierarchy collapses at some level k, meaning that the class of depth k is equal to the class of depth $k + 1$ and therefore to all classes of depth j with $j > k$. Sipser (1983) proved that these hierarchies are proper for the neuron type [AND, OR] and the neuron type [AND, OR, NOT]. For the neuron type [AND] and the neuron type [PARITY], in contrast, these hierarchies collapse already at level 1 because all neural circuits using only AND neurons or only PARITY neurons can be flattened to just one neuron. Unfortunately, whether the hierarchies for the more interesting types (THRESHOLD or MAJORITY) are proper is still an open question.

Another interesting question is whether some neuron types are equivalent to or significantly more powerful than other types. It is well-known that each neuron of type AND can be replaced by a neural circuit of depth 3 consisting of neurons of type [OR, NOT] and, conversely, neurons of type OR can be replaced by a neural circuit of depth 3 consisting of neurons of type [AND, NOT] (e.g., x *and* y can be replaced by *not(not(x) or not(y)))*. Therefore we may conclude that the neuron type [AND, OR, NOT] is equivalent to the neuron type [AND, NOT] (and to [OR, NOT]) because the depth of corresponding circuits differs only by a constant factor. The same is true for the neuron type THRESHOLD and MAJORITY because, on the one hand, a MAJORITY neuron is a special case of the THRESHOLD neuron with threshold *fan-in/2* and, on the other hand, we can replace each THRESHOLD$^\theta$ neuron with *fan-in* n by a MAJORITY neuron by doubling the number of inputs (now the threshold of the MAJORITY neuron equals n) and connecting $n - \theta$ of the additional inputs to a constant input of *1* and the rest to a constant input of *0*.

Obviously, THRESHOLD is at least as powerful as AND and OR because they are special cases with threshold = *fan-in* and threshold = 1, respectively. Furst et al. (1984) have proven that THRESHOLD is significantly more powerful than [AND, OR, NOT], because the Boolean function parity can be computed by a neural circuit of THRESHOLD neurons in depth 2 with a hidden layer of n neurons (for input dimension n), whereas parity cannot be computed by neural circuits of neuron type [AND, OR, NOT] for any constant depth and polynomial size bound (reminder: of course, parity, like any arbitrary Boolean function, can be computed in depth 3 using exponentially many neurons of type [AND, OR, NOT]).

Because the first formal neuron model, the McCulloch-Pitts neuron, is also equivalent to the THRESHOLD and MAJORITY neuron, we may summarize that these neuron types are equivalent with respect to the complexity class of constant depth and polynomial size, but significantly more powerful than neurons of type [AND, OR, NOT]).

We may also generalize our model of the neural circuit from one binary output neuron to many binary output neurons. Chandra et al. (1984) showed that addition can be computed in constant depth by neurons of type [AND, OR, NOT]). They constructed a neural circuit of size $O\left(n^2\right)$ and depth 3 for the addition of two binary numbers (input dimension $n + n$, output dimension $n + 1$). For the computation of multiplication and multiple addition with polynomial size and constant depth we need THRESHOLD neurons. Parberry (1994) constructed a neural circuit of size $O\left(n^2\right)$ and depth 6 for the addition of n binary numbers of length n and thereby we get as a simple corollary a neural circuit for the multiplication of two binary numbers of length n with size $O\left(n^2\right)$ and depth 7.

Another obvious generalization for THRESHOLD circuits would be to enhance the neuron model by allowing connections with variable weights, i.e., the neuron no longer simply counts the number of active inputs but instead computes a weighted sum of input values: output 1 if $\Sigma\, w_i \cdot input_i \geq \theta$. This seems biologically plausible because the strengths of the synapses (i.e., the coupling strengths of the neurons) differ and this is modeled by the variable weights. The size of the weights can be limited to an exponential bound because Muroga et al. (1961) proved that whenever a Boolean function can be computed by a THRESHOLD neuron with weights, then there exists such a neuron with weights smaller than $(n + 1)^{(n+3)/2}/2n$. Conversely, there are Boolean functions which require exponential weights: Hastad (1992) constructed a Boolean function computable by a THRESHOLD neuron with weights, but requiring a maximal weight greater than $n^{n/(2+\varepsilon)}$ for any $\varepsilon > 0$, which comes quite close to the upper bound $((n + 1)^{(n+3)/2}/2n < (n + 1)^{n/2})$.

The necessity of exponential weights seems biologically questionable, because it requires neurons of very high precision and bandwidth. Fortunately, they are not needed: because the length of the binary representation of the weights can be limited to $\log((n + 1)^{(n+3)/2}/2^n) = O(n \log n)$, we can replace each THRESHOLD neuron with weights by a neural circuit of simple THRESHOLD neurons with size $O(n \log n)^2$ and depth 7 by using again the neural circuit for multiple addition (to sum up the weights with input 1). This means that we can transform any neural circuit of THRESHOLD neurons with weights into one without weights and thereby increase the depth at most by factor 7, i.e., the property of constant depth remains invariant. Goldmann et al. (1992) even improved this result, proving that a THRESHOLD circuit with weights can be transformed into one without weights and with the depth increased only by two.

A more modest and more biologically plausible generalization would be to enhance the THRESHOLD neuron with weights of at most polynomial size. This is equivalent to using multiple connections between two THRESHOLD neurons (without weights), where the number of connections is also polynomially bounded by the polynomial bound for the network size. The corresponding complexity class of neural circuits of polynomial size and depth k using THRESHOLD neurons with weights of polynomial size is called TC_k^0 (note that any function in TC_k^0 can be computed by a neural circuit of simple THRESHOLD neurons in depth $k + 1$ by substituting multiple connections via multiple neurons). According to the above considerations it is an open question whether the classes TC_k^0 form a proper hierarchy because we can substitute a circuit of simple THRESHOLD neurons for the neurons with weights and we already noted that the hierarchy question is an open problem for the THRESHOLD neurons. Hajnal et al.

(1987) proved that the inclusion of TC_2^0 in TC_3^0 is proper: They proved that the computation of the *inner-product-mod-2* is in TC_3^0 but not in TC_2^0. Even if it is generally believed that the hierarchy is proper, it is still an open problem whether the hierarchy collapses into TC_3^0.

Another interesting generalization is to use neurons with a continuous valued, nonlinear (e.g., sigmoid) output function instead of the step-like threshold function. Maass et al. (1991) proved for polynomially bounded weights that such THRESHOLD networks using neurons with sigmoid output functions can be simulated by THRESHOLD networks with binary output values with no increase in the depth of the circuit (and, of course, vice versa). Therefore, all results for THRESHOLD networks with polynomially bounded weights also hold for sigmoid output functions (as used in back-propagation networks). The same may not be true for exponential weights. Sontag (1992) constructed a differentiable nonlinear output function such that any Boolean function can be computed by a THRESHOLD network using just two neurons with sufficiently large weights.

At this point, it is interesting to reconsider the old question of how many hidden layers are needed for neural circuits with THRESHOLD neurons or, even more interesting, but similar, for the well-known multilayer perceptron. Trivially, any Boolean function can be computed with one hidden layer (depth 2) using exponentially many THRESHOLD neurons with polynomially bounded weights. Hajnal et al. (1989) proved that at least two hidden layers (depth 3) are needed if the circuit size is polynomially limited. Whether two hidden layers suffice remains an open problem.

The analogous hierarchy of neural circuits with neurons of type [AND, OR, NOT] is called AC_k^0. We already noted that this hierarchy is proper. Allender (1989) proved that each Boolean function of this hierarchy is computable by a THRESHOLD circuit of depth 3 and size $O\left(n^{(\log n)^c}\right)$ for a constant c. Therefore, it seems plausible that the whole hierarchy is a subset of TC_3^0 (this subset would be proper because parity is in TC_2^0, but not in AC_k^0 for any k).

Summarizing, we may state for threshold circuits with constant depth and polynomial size:

- The neuron types McCulloch-Pitts, MAJORITY and THRESHOLD are equivalent.

- THRESHOLD neurons with and without weights are also equivalent; corresponding networks differ only by a factor of two in depth (weights of polynomially size seem biologically plausible).

- THRESHOLD neurons are significantly more powerful than neurons of type [AND, OR, NOT] (parity is only computable with THRESHOLD neurons).

- It is an open question whether two hidden layers suffice for THRESH-OLD circuits.
- It is plausible that two hidden layers of THRESHOLD neurons with weights of polynomial size suffice for computing any Boolean function computable by [AND, OR, NOT] circuits in constant depth and polynomial size.

3.3 Neural Circuits With Logarithmic Depth

The most simple neurons have one or two inputs and one output. Examples are the NOT, IDENTITY, CONSTANT_0, CONSTANT_1 with one input and AND, OR, XOR, NAND, NOR, etc., with two inputs. There are only 4 different neurons with one input and 16 different neurons with 2 inputs. If the output depends on the whole input, we need at least a neural circuit of depth $\log n$ (e.g., AND $(x_1, x_2, ..., x_8,) = (x_1 \text{AND } x_2)$ AND $(x_3 \text{ AND } x_4)$ AND $((x_5 \text{ AND } x_6)$ AND $(x_7 \text{ AND } x_8))$ which has depth $3 = \log 8$). Therefore, we may ask what kinds of Boolean functions can be computed by neural circuits of neurons with only two inputs (i.e., *fan-in 2*) in logarithmic depth. This class is called NC^1 (Nick's class, in honor of Nick Pippenger). Of course, we do not need the whole set of neurons of *fan-in* ≤ 2. Any basic set, e.g., [AND, NOT], [OR, NOT] or [NAND] would suffice because we could replace the other neurons by a small circuit of each basic set.

Hong (1986, 1987) proved that multiple addition is in NC^1 by constructing a neural circuit of size $0 (n^2)$ and depth $0 (\log n)$ for the addition of n binary numbers of length n. From this result, we can easily construct a neural circuit for the THRESHOLD neuron, i.e., THRESHOLD is in NC^1. As a further consequence we get that $TC^0 = \cup TC_k^0$ is a subset of NC^1, because we can replace in any THRESHOLD circuit of constant depth every THRESHOLD neuron by the neural circuit of depth $0 (\log n)$ using neurons with *fan-in 2* and thereby get a neural circuit of depth $0 (\log n)$ using neurons with *fan-in 2*. If we define $AC^0 = \cup AC_k^0$ we can summarize: $AC^0 \subset TC^0 \subseteq NC^1$ where the first inclusion is proper.

As already noted, logarithmic depth is also compatible with the 100 step rule. Therefore, we could define THRESHOLD circuits of logarithmic depth. This class is called TC^1. If we define AC^1 for neural circuits of neuron type [AND, OR, NOT] and logarithmic depth we get: $NC^1 \subseteq AC^1 \subseteq TC^1$. An obvious generalization is to consider neural circuits of polylogarithmic depth. NC^k is defined as the class of Boolean functions computable by neural circuits of depth $0((\log n)^k)$ using neurons with *fan-in 2*. Accordingly, we define AC^k and TC^k and get: $\forall k \geq 0 : AC^k \subseteq TC^k \subseteq$

$NC^{k+1} \subseteq AC^{k+1}$. Unfortunately, we know only that the first inclusion, $AC^0 \subset TC^0$, is proper. Moreover, if we consider the class of Boolean functions computable by neural circuits of any polylogarithmic depth, then all three classes coincide: $NC = \cup NC^k$, $AC = \cup AC^k$ and $TC = \cup TC^k$ are all equal to each other, but it remains an open problem whether all three hierarchies collapse already into TC_3^0, that is, whether two hidden layers of THRESHOLD neurons might suffice.

Summarizing our considerations on neural circuit complexity theory, we may conclude that THRESHOLD circuits with constant depth and polynomial size are powerful computing models (TC^0). The use of polynomial weights is not necessary, but can simplify the neural circuit. In particular, the high precision and bandwidth of exponential weights is not required. If instead we use neurons with *fan-in 2* (or any other constant bound for the *fan-in*), we need logarithmic depth. The computing strength of such neural circuits (NC^1) is at least as strong as THRESHOLD circuits of logarithmic depth ($TC^0 \subseteq NC^1$). In other words, we can choose between constant *fan-in* with only logarithmically bounded depth (time complexity) and unbounded *fan-in* with constant depth. On the other hand, THRESHOLD circuits (with unbounded *fan-in*) are significantly more powerful than [AND, OR, NOT] circuits in constant depth. Whether there are more powerful types of Boolean neurons than THRESHOLD is an open problem, because $AC = TC = NC$ might collapse to TC_3^0: If collapse (to any constant depth) occurs, no Boolean function in NC ($= AC = TC$) is significantly more powerful than THRESHOLD.

3.4 *Neural Circuits With Loops*

Neural circuits without loops (internal feedback connections) are computing devices without memory. The output of the circuit depends only on the current input but not on former inputs. This is no problem as long as all necessary information for computing the output is available in the input; otherwise, memory is necessary. For example, if we read a book, we focus on syllables at each time step. In order to understand the meaning of the written sentences, we have to remember the written context.

The simplest form of memory may be located directly in the input neurons s^i. This is true if the input neurons compute a moving average, as a capacitor in an electrical circuit does:

$$s^i(t+1) = \alpha \cdot s^i(t) + (1 - \alpha) \cdot inp(t+1) \text{ with } \alpha \in (0,1) .$$

In this equation, the actual state of the neuron $s^i(t+1)$ depends on its former state $s^i(t)$.

Another example is normalization via feedback within a layer of neurons. For example, our optical sensors can adapt to a broad range of light intensities. We can detect edges in pictures by using relative differences in local light intensities: Edges are characterized by a large discontinuity in light intensity.

Feedback from higher levels of representations in the neural circuit also may be important. For example, the error rate in letter recognition is decreased when we can use the context of the text already read to limit the search space (e.g., a *u* must follow a *q*, *b* cannot follow *po*, etc.).

Finally, feedback can be used for repeating procedures. For example, if we have an imperfect neural circuit for noise cancellation or pattern completion, the iterated execution of this procedure would improve performance. This can be done by feeding the output back to the input.

Therefore, we may state that feedback loops offer some advantages for realizing prerational intelligence. A disadvantage is the halting problem: Does the computation terminate? For example, the neuron s with the update rule $s(t+1) = -s(t)$ does not terminate when instantiated with 1; instead, its state varies with a period of length 2. In general, the computation has to be periodical if the states of the neurons are discrete (finite range) and the update rule of the neurons is deterministic. The reason is that there are only finitely many different states of the whole neural circuit – although the number is exponential in the number of neurons – so one or more states must be repeated at some point, making the computation periodical relative to the first occurrence of this state.

In his seminal paper, Hopfield (1982) investigated a THRESHOLD circuit with feedback loops. In honor of him it is called a *Hopfield Network*. He constructed a so-called energy function for which he proved that this function is monotonically decreasing if each neuron is asynchronously (one at a time) updated and the connection strengths between two neurons are symmetric. More precisely, this energy function is a quadratic polynomial and each neuron changes its state at updating if and only if the energy decreases (in the special case that the energy remains constant, then the neuron is assigned by definition the default value 1). For synchronous updating (i.e., all neurons are updated at the same time), the energy function does not necessarily decrease, but the network will converge to a cycle of length 2 (as a special case, the network may converge to a single stable state).

Unfortunately, the number of update steps until convergence can be quite long. The convergence time can be polynomially bounded in the size of the weights of the THRESHOLD neurons because in this case, the energy function is polynomially bounded, i.e., there are only polynomially many different values of the energy function. Fogelman et al. (1983) proved in a more careful analysis that the convergence time can be bounded by $0(n^2 w)$ where w is the maximal weight of the network and n the number of neurons. Of course, there are at most 2^n different states of a neural circuit with n neurons, therefore, the convergence time has to be smaller than 2^n. But this bound cannot be significantly decreased because there are ("pathological") examples of THRESHOLD circuits (using exponential weights) with exponential convergence time. This was proven for synchronous updates by Goles and Olivos (1981), for any synchronous update rule by Haken (1989) (for a special asynchronous update rule, see also Haken 1988).

Nevertheless, in practical applications of Hopfield networks for noise reduction and pattern completion convergence is achieved within a few synchronous or asynchronous updates of all neurons (e.g., less than 10 updates). Therefore, we could also stop the updating after a fixed amount of time and read out the final network state for further information processing; in this way we can obey safely the 100 step rule.

Kleene proved already in 1956 that THRESHOLD circuits with feedback loops are equivalent to a finite state machine. More precisely, any finite state machine can be efficiently simulated by a THRESHOLD circuit of the same size. The same is true for neural circuits with neurons of type [AND, OR, NOT] with unbounded *fan-in*.

To simulate Turing machines, we need space for representing the potentially infinite memory (i.e., for any given input we need only finite memory but the size of the required memory space depends on the input). There are two possibilities (Hartley & Szu 1987): Either we use potentially infinite networks on the one hand, or finite networks with real-valued neurons of arbitrary precision on the other. Sigelmann and Sontag (1992) improved the construction for the latter possibility by simulating Turing machines rather efficiently: The required precision grows only linearly in the memory requirement of the simulated Turing machine.

It should be remarked that simulating Turing machines with potentially infinite neural circuits is not very satisfying because we simulate the passive memory with active processing elements (neurons), or, in other words, we do not use the parallel power of neural circuits but misuse most of the neurons just for storing bits. The simulation of finite state machines also does not use the parallel power efficiently: In every step only one neuron is active. Of course, these facts are not very astonishing because we try to simulate sequential machines with massively parallel machines, which has to be "overkill".

4. PROBLEM SOLVING

In the following, we want to discuss the basic problem-solving paradigms of computer science and how these could be implemented in neural networks realizing prerational intelligence.

Problem solving can be viewed as both a search problem looking for a solution and, more generally, an optimization problem looking for the best solution. Each approach can be transformed into the other: A search problem is an optimization problem where the solutions have value 1 and all others value 0. Vice versa, an optimization problem can be reduced to the search problem: *Find a solution with value better than c.* We can solve the optimization problem using a binary search for the best value of c.

Moreover, optimization can mean minimization or maximization of a given optimization function. Both problems are equivalent, because the maximization of a function f is the minimization of the negated function $-f$.

Basic paradigms for optimization are *hillclimbing, gradient descent,* and *evolution* whereas those for search are *divide-and-conquer* and *branch-and-bound.* Of course, according to the above remarks we may use search strategies for reducing the search space of the corresponding formulation as an optimization problem and, conversely, we may use optimization strategies for solving a search problem by optimizing the estimated distance to a solution. Further solution strategies are *recursion* and *dynamic programming*.

4.1 *Hillclimbing*

The metaphor of hillclimbing is a "blind" individual searching for the highest point in the optimization landscape by accepting each trial step if and only if it leads to a higher point. A step is a small variation of the actual standpoint in the search space. The individual has no memory, that is, the direction of the trial does not depend on the history.

A neural network implementation of this strategy is the Hopfield Network. Each neuron changes its state if and only if the corresponding energy function decreases (or remains constant for a change to value 1), or, in other words, if the negative energy function increases (*hillclimbing*). Therefore, the Hopfield Network always converges to a local minimum of the energy function, that is, one in which no change of state in a single neuron decreases the energy function.

The energy function is a quadratic polynomial. Conversely, we can construct for any quadratic polynomial f a Hopfield Network with energy function f. Therefore, the Hopfield Network solves the minimization problem for quadratic polynomials. Of course, it finds only a *local* minimum and not the *global* minimum. The problem of finding the global minimum for quadratic polynomials is NP-complete in the n-dimensional domain of integer or binary vectors; that is, there exists no polynomial algorithm unless the complexity classes P and NP are equal. In other words, there is no efficient (polynomial) algorithm available which is guaranteed to find the global minimum, not merely a local minimum. Another consequence of the NP-completeness property is that all problems in NP, that is, nearly all optimization problems of practical relevance can be reduced to this problem.

Summarizing we may conclude that the Hopfield Network realizes a hill-climbing strategy for a rather general optimization problem. However, the quality of the achieved solutions depends on the optimization problem. For example, the known results for the traveling salesman problem are very poor (Braun 1991). Nevertheless, some problems solved by prerational systems appear suited to such hillclimbing strategies. An example is the correspondence problem, i.e., determining what pixels in two stereographic images correspond (Braun 1990). This problem has to be solved by the visual system in computing the spatial depth of viewed objects.

4.2 *Gradient-Descent*

The metaphor of gradient descent is a "short-sighted" individual searching for the lowest point in the optimization landscape by walking in the direction of the steepest descent. The length of the step may be fixed or depend on the steepness (length of the gradient vector). The individual also has no memory, that is, the search direction depends only on the actual gradient, not on the history.

A neural network implementation of this strategy for the minimization of quadratic polynomials is the continuous Hopfield Network (i.e., THRESH-OLD neurons with sigmoid output functions instead of step-like output functions) (Hopfield 1995) and the brain-state-in-a-box model from Anderson (1977) (i.e., THRESHOLD neurons with truncated linear output functions). These approaches are not only suited for real-valued minimization but also for minimization in the domain of binary vectors. We just interpret the output value of the neuron probabilistically: An output value $p \in [0, 1]$ means that the corresponding component of the solution vector is probably 1 with probability p (or 0 with probability $1 - p$). By this generalization, we can impressively improve the quality of the achieved solutions. The reason is that we need not make discrete decisions (represented by binary values of the neurons) in the beginning of the search, but we can navigate with fuzzy decisions through the search space, favoring each component more or less strongly until in the end the values of the neurons become fixed in a corner of the hypercube. Because the output values are bounded by $[0, 1]$, Anderson called his model brain-state-in-a-box. More precisely it realizes a projected gradient descent; whenever the state of the neural circuit is on a limiting hyperplane of the search space, the gradient is projected onto this hyperplane in such a way that the hypercube (search space $[0, 1]^n$) cannot be left. For a detailed discussion of these models see also Braun (1991).

Summarizing, we may state that THRESHOLD neurons with real valued output and feedback loops (like the continuous Hopfield model and the brain-state-in-a-box model) can realize gradient descent. Moreover, this approach achieves even for discrete-valued optimization problems significantly better performance than the hillclimbing approach using THRESHOLD neurons with binary output (Braun 1990, 1997).

4.3 *Evolution*

Evolutionary algorithms are optimization strategies simulating biological evolution. The optimization function is called fitness. The population is a set of possible solutions. They represent the memory trace of the prior search or the possible starting points for further search. Parents with above average fitness are preferred for generating offspring (*selection*). Two commonly used operators for producing offspring are *mutation* and *crossover*. Mutation constructs an offspring by making a small change in the properties of one parent (asexual reproduction) whereas crossover mixes the properties of two parents (bisexual reproduction). The size of the population is kept constant by keeping the best individuals (*survival of the fittest*) and replacing the less fit individuals with the new offspring.

Because nature uses evolutionary strategies to construct and optimize neural networks in combination with learning, it is not very astonishing that artificial evolution based on fast learning algorithms is also an efficient tool for optimizing artificial neural networks (Braun 1994). In other words, the paradigm of evolution is good for constructing circuits to realize prerational intelligence. On the other hand, because evolution is a long-term process, it seems not to be suited for implementing prerational intelligence itself (i.e., for realizing the neural algorithms to compute individual solutions).

4.4 *Divide-and-Conquer*

The idea of divide-and-conquer is to *divide* a problem into simpler subproblems and then to solve (*conquer*) these separately. One famous example of this strategy is the polynomial algorithm for solving the optimization problem of linear programming, that is, minimizing linear functions with linear constraints (inequalities). A special case is the loading problem for the perceptron (THRESHOLD neuron with real valued inputs), that is, finding the weights such that the output of the perceptron is 1 for all given positive examples and 0 for all given negative examples. In every step, the polynomial search algorithm of Karmarkar halves the search space and determines in which half a solution must lie if there is one (Kamarkar 1984).

Another example is our daily distribution of work. We educate human experts or construct mechanical experts to be specially tailored for *conquering* subtasks. The whole task is *divided* into subtasks and distributed to the appropriate experts. To implement this heuristic, we need on the one hand an algorithm to classify the given input for assignment to the appropriate expert and on the other hand algorithms for each expert to solve the subtasks. For the first part we can use any neural classifier model such as learning vector quantization, self-organizing feature maps (Kohonen 1982), or adaptive resonance theory (Grossberg 1984). Each class i (= union of tasks for expert i) is represented by an output neuron which fires if and only if the task given in the input should be distributed to expert i. Ritter used self-organizing feature maps for classifying the experts and simple linear functions as experts which solved the local problem using a Taylor approximation of degree 1 (Ritter et al. 1990).

4.5 *Branch-and-Bound*

The branch-and-bound method is used for solving search problems in discrete domains by efficiently traversing a search tree. We will explain this method for the domain of binary vectors $[0, 1]^n$. A corresponding search tree is built by connecting each binary vector $x \in [0, 1]^k$ with its offspring $x0, x1 \in [0, 1]^{k+1}$. This means that the subbranches of branch 010 are all binary vectors beginning with 010 and, in particular, the leaves of this branch are all binary vectors of the domain beginning with 010. In order to get a first solution, we descend the search tree from the root using the most promising branch. Then we complete the search by navigating through the tree according to the following strategy:

- whenever we can determine for a subbranch (without actually descending along it) that the best solution in this branch has to be worse than the previous best solution (*bound*) we cut off this branch
- if there are subbranches left, we descend to the most promising,
- if it is a leaf we test for improving the bound (best solution),
- otherwise, we ascend back to the next predecessor with branches that have not yet been considered.

The critical ingredients of this strategy are good heuristics for selecting *most promising subbranches* and for determining bounds for the best solution of a subbranch. Without such heuristics, we have to navigate through the whole search space before the algorithm stops. This exhaustive search is exorbitantly time consuming for standard optimization problems with typically exponential search spaces. Nevertheless, Grötschel and Holland (1991) and Padberg and Rinaldi (1987) used this strategy to solve very large traveling salesman problems with up to a thousand cities (remember that presumably no polynomial algorithm exists because the traveling salesman problem is NP-complete).

A typical application of such an algorithm is used for programming chess playing machines which try to search efficiently for the best solution in depth k. This strategy may also be used by a human chess (or other game) player when he rationally analyzes a given board position. For this reason, I would argue that the branch-and-bound method should be classified as a method for rational thinking, not for prerational intelligence. Moreover, navigation through the search tree typically is very time consuming, so that the 100 step rule (our necessary condition for prerational intelligence) will be greatly exceeded.

4.6 *Recursion*

The recursion operator was already introduced in Section 1. The idea can be characterized as follows: Not only are we allowed to divide the given task into simpler subtasks (see divide-and-conquer), but we may also use a hopefully simpler instantiation of the very same task. For example, the $n - th$ Fibonacci number can be computed by the following recursive formula: $f(n) = f(n-1) + f(n-2)$, or multiplication by the following recursion: MULTIPLY (x, y) = MULTIPLY $(x, y + 1) + x$. We label the recursion primitive (see Section 2.2) if the recursive call of the same task gets a smaller input, as in both the above examples. In this case, the algorithm has to terminate because the task gets strictly simpler at every recursive call (application of the divide-and-conquer method).

An example of general recursion is game playing: Given a function f which computes to a board position x, the succeeding position $f(x)$ after one's own move and that of the opponent. Then we can reduce the iterative steps of game playing to the recursive function: $g(x) = x$ if x is already a final board position and $g(x) = g(f(x))$ else. Here, we cannot say that the board positions get simpler. Nevertheless, we know that some games always terminate (e.g., backgammon, because the stones always move in one direction), but for other games we have to limit the number of moves (e.g., for chess the number of moves without removing a piece is limited to 50). In general, the termination problem is not decidable for general recursion (μ-recursive operator).

Recursion can be easily implemented in neural networks with feedback loops. According to our 100 step rule as a necessary condition for prerational intelligence, we may state that general recursion is too time consuming. Even primitive recursion has to be limited to a constant time bound. Examples of such limited recursion through feedback loops are pattern completion and noise reduction with Hopfield Networks: In practical evaluations, results are sufficient when the recursive feedback loop is limited to a few cycles (e.g., less than 10). A more sophisticated example is language processing where lower and higher levels of language processing (e.g., circuits handling phonemes, words or meaning, respectively) cooperate in both directions.

4.7 Dynamic Programming

Dynamic programming is a method for optimizing a sequential decision problem: The task is to find the cheapest path of length k from a given starting point A, given a finite graph G with weighted edges (the weights are the costs for using the edge) and weighted nodes v_A (these weights represent the value of a final node). The cost of a path is the sum of the weights of the edges used plus the value of the final point. The method relies on the Bellman principle: The cheapest path from A of length k can be partitioned into the edge from A to some neighbor B and the cheapest path from B of length $k - 1$. The method of dynamic programming can be summarized as follows. We generalize the problem to D^k: Find for *all* possible vertices A of graph G the cost $V^k(A)$ of the cheapest path from A using k edges.

The generalized problems are solved sequentially: Having solved D^k, we can solve D^{k+1} by the Bellmann principle:

$$V^0(A) \quad := \quad v_A, \text{ with value of node } A = v_A$$

$$V^{k+1}(A) \quad := \quad \min\{c_{AB} + V^k(B) \mid (A, B) \text{ is an edge in } G\},$$

$$\text{with weight of edge } (A, B) = c_{AB}.$$

With solutions for the generalized problems D^j, that is, knowing the cheapest path of length j for all vertices and all $j \leq k$, we can easily solve the specific problem and construct the cheapest path $(x_0, x_1, ..., x_k)$ from vertex A:

$x_0 = A$,

x_{i+1} *is determined by*

$V^{k-1}(x_{i+1}) := \min\{c_{x_{i+1},x_i} + V^{k-i-1}(x_{i+1}) \mid (x_{i+1}, x_i) \text{ is an edge in G}\}$.

It seems astonishing that the generalization of the problem speeds up the solution, but the following simple consideration shows that otherwise we have to test exponentially many possible paths starting from A whereas dynamic programming is a polynomial algorithm with time complexity $O(kn^2)$: Each of the k steps of the algorithm uses at most $O(n^2)$ time steps, because there are n vertices and at most n neighbors for each vertex.

Dynamic programming could also be generalized for stochastic scenarios:

1. The next state depends stochastically on the chosen action instead of deterministically on the choice of an edge.
2. The costs of an action (or the reward of the subsequent state) is stochastic.

In other words, we are looking for a path with minimal *expected* costs.

Moreover, dynamic programming can be generalized for infinitely long paths by using an exponential decay factor γ $(\gamma < 1)$ for summing up the costs of the edges used, that is, the cost of the $k - th$ edge is weighted by k. Formally,

$$V^k(x_0) = \min\left\{\sum_{i=0}^{k} \gamma^k \cdot c_{x_i x_{i+1}} \mid (x_0, x_1, \ldots, x_k) \text{ is a path from } x_0\right\}.$$

In this case, the Bellman principle has to be transformed by weighting the cost of the best path from the neighbor accordingly with factor γ:

$$V^{k+1}(A) := \min\left\{c_{AB} + \gamma \cdot V^k(B) \mid (A, B) \text{ is an edge in } G\right\}.$$

Because of the exponential decay factor $V^k(A)$ converges and the value $V(A)$ of the infinitely long cheapest path is well defined by:

$$V(A) \; : \; =$$

$$\lim_k V^k(A) = \min\left\{\sum_{i=0}^{\infty} \gamma^k \cdot x_{x_i, x_{i+1}} \mid (x_0, x_1, x_2, \ldots) \atop \text{is an infinite path from } x_0\right\}.$$

There are many important examples of the efficient application of dynamic programming. For example, the shortest path problem (from a given starting point A to a goal B in a weighted graph G, which is the formalization of our daily problem: find the shortest way from the actual standpoint to a target place given a roadmap) can be interpreted as a dynamic programming problem by adding self-reflexive edges of weight 0 from goal B to itself, in order to allow remaining at the goal, and by penalizing with maximal costs all final points different from goal B. Another example is the matrix multiplication problem: how to apply optimally the associative rule for computing the product of n matrices.

An important application in the scope of prerational intelligence is the reinforcement learning problem, i.e., the optimal strategy to choose sequentially n actions (edges) such that the sum of the received reinforcement signals (i.e., costs of the edges) is optimized. Of course, the number of states (i.e., vertices in G) may be exorbitantly high in practical applications; the extreme is the problem which each animal faces in finding its optimal way through life. Nevertheless, through generalization from only a few examples this problem can be handled efficiently: The method of temporal difference learning (see Section 6, and Ritter & Bauer, this volume) is an approximation of dynamic programming and a fundamental method for constructing or optimizing neural networks for sequential decision problems (Sutton 1988). Tesauro (1992) proved for an artificial sequential decision problem that this approach can lead to a sophisticated solution: A multilayer perceptron trained with temporal difference learning to play backgammon performed at the level of the ten best players in the world.

Therefore, we may conclude that temporal difference learning or similar approaches (generally called reinforcement learning) are strong methods for constructing neural networks to solve sequential decision problems, a fundamental task for prerational intelligence. Of course, these approaches cannot guarantee an optimal solution because they are only approximations of dynamic programming (which is guaranteed to be optimal). On the other hand, we cannot hope for more because the reinforcement problem is NP-hard when the number of states (i.e., vertices in G) may be exponential.

5. CONSTRUCTING NEURAL NETWORKS

For developing large software systems the leading principles are

- *Modularization*: Large software systems are composed of several modules which communicate through interfaces (input and output channels of the modules).
- *Information hiding*: The modules hide their internal information processing. Particularly, in object-oriented programming the modules offer methods with asserted functionality. Other modules can use these methods, but their implementation is hidden, only their functionality is guaranteed. The advantage of this approach is the software evolvabil-

ity. The implementation of each module can be exchanged (improved), independent of the other modules ("plug in and play").

- *Hierarchies of information processing*: There are two hierarchies of information processing – semantic and structural. The structural hierarchy means that the complete software system is composed of modules. These large modules may be also composed of smaller modules. The semantic hierarchy corresponds to the information processing: Consider language processing for example. On the first stage we detect frequencies in a stream of acoustic input. These are classified as phonemes. Consecutive phonemes are classified as words. Consecutive words are assigned to propositions (meaning). For speech production, we have to proceed in the opposite direction: From meaning to frequencies.

In biological information processing systems (neural networks), we can identify the same principles. Large modules are the different compartments of the brain which are themselves composed of modules. In software systems, modules are defined by a coherent part of the program text whereas in neural systems modules are a coherent part of the neural network, i.e., they form a neuron cluster without overlap to other modules. Another characteristic is the interface. The neural network modules are sparsely connected to the other modules by few connections (axons) whereas the internal structure is strongly connected.

In contrast to software systems, the neural network modules consist of components (neurons or neuron clusters) with similar functionality. In software systems we implement just one expert (solver) for each subproblem (module) whereas the neural network modules consist of a committee of experts which vote for their common output like in a democracy (see mixture of experts, Jordan & Jacobs 1994) and local linear maps (Ritter et al. 1990, ch. 13): This vote is the typical function of an artificial neuron like the PERCEPTRON in multilayer networks or THRESHOLD neuron in circuit complexity theory.

This democratic principle has several advantages.

- *Robustness* (graceful degradation): The malfunction of a part of the network only degrades but does not destroy the performance.
- *Parallelization*: Each neural expert can inpendently compute its function and communicate only the result.
- *Evolvability*: The function of neural networks can be changed by mutations in small steps with only small changes in the performance (stochastically: the offspring could be better or worse). Examples for mutations would be deleting an expert or adding a slightly changed copy of an expert.

- *Learning*: The performance can be improved by small changes of the parameters (i.e., weights of the neurons) of each neural expert. An example is minimizing the error by gradient descent.

These are the major differences to software systems. Small changes (e.g. inverting a single bit) can already stop a software system. Human developers try to minimize their lines of source code and therefore implement just one expert for each module. Only for critical modules do they duplicate the hardware and the software. Nevertheless, duplication does not achieve graceful degradation. For example, if we duplicate two software systems, then we may achieve fault tolerance for a single failure, but already two small failures in both systems can cause a crash. Duplication does drastically reduce the probability of a crash for reliable systems. If the probability of a failure in one system is p, then we may conclude that the probability of a failure in both systems is only p^2.

Of course, we may model biological neural networks as artificial neural networks. These models also follow the above democratic principles and are useful for modeling prerational intelligence. The specification of artificial neural networks consists of two parts: the network topology (number of neurons and their connections) and the parameters (weights) of each neuron. The techniques for tuning the weights can be summarized as learning and are discussed in the next section, whereas the techniques for the specification of the topology can be summarized as evolution. This can be done manually: The designer tests a set of different topologies by trial and error and then refines the best by adding or deleting neurons. This selection can also be done automatically by using an evolutionary algorithm. Moreover, there are two important special cases with population size 1:

- *Constructive algorithms*: The topology is enhanced in each step by a few neurons and connections. This process is stopped, when no further improvement of the performance is measured. An example is cascade correlation (Fahlman & Lebiere 1990; Littmann & Ritter 2000).

- *Destructive algorithms*: Starting with a topology that is clearly too large, neurons and connections are incrementally removed until the performance deteriorates. The selection of the removed neurons follows some heuristic principles like "remove the connection with the smallest weight". Examples are pruning, weight elimination, optimal brain damage, optimal brain surgeon (Hassibi & Storck 1993; Stahlberger & Riedmiller 1997).

A more detailed overview of topology optimization can be found in Braun (1997, 1999).

6. STRATEGY LEARNING

A fundamental task for every individual is to learn strategies for optimizing its behavior in a given environment. In Figure 1, we illustrate the general situation. The agent computes an action a according to its strategy $S : a = S(x)$. After each time step the new situation x^{t+1} depends on the reaction R of the environment or system, respectively: $x^{t+1} = R(a, x^t) = R(S(x^t) x^t)$.

We can distinguish three different cases of learning a strategy:

1. *Supervised Learning*: There exists an expert who already has an appropriate strategy and teaches the agent by giving a set of examples: in situation x choose action a.

2. *Goal directed learning*: The teacher possesses no strategy but provides goals to be achieved.

3. *Reinforcement learning*: There is a reinforcer who penalizes bad situations and rewards good situations. The task is to optimize the expected reinforcement signals.

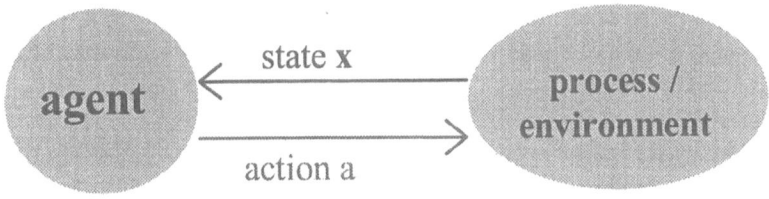

Figure 1: Interaction cycle between the agent and its environment.

From the first case to the last the necessary domain knowledge decreases whereas the task for the learner gets harder. More precisely, in the first case we need an omniscient supervisor (expert), whereas in the last case the learner explores the consequences of his actions by himself. In the following, we give an overview of techniques used for solving these problems in the domain of artificial neural networks, which is taken as the model for systems exhibiting prerational intelligence. However, we do not claim that the same algorithms are used in biological brains, but point out that the elucidation of biological learning algorithms at the neural level is very hard and, to date, has only been worked out in part for the simple Hebbian learning rule.

6.1 *Supervised Learning*

The task for the learner in supervised learning is to imitate his teacher. Typically, the domain (the set of all possible situations x) is by far too large to be

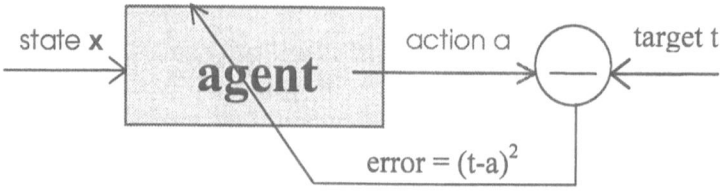

Figure 2: Learning cycle for supervised learning with example x and target t.

completely included in the learning set. Instead, the teacher gives prototypi-cal examples which have to be generalized by the learner to the complete do-main. According to the motto, "make theories as simple as possible but not simpler" (see Occam's razor), the learner tries to find the simplest function which can fit the given learning set. The available set of functions depends on the agent. In conventional artificial intelligence approaches the learner typically uses a rule system (also called an expert system), whereas in arti-ficial neural network approaches the learner typically uses a THRESHOLD circuit. The most common method is the multilayer perceptron in which the learner uses gradient descent (also called backpropagation) to minimize the error of a given feedforward THRESHOLD circuit with continuous weights and sigmoid output functions. The overall error is the sum over all the indi-vidual errors for each learning example which are defined as the square of the difference between the actual output and the desired output given by the teacher (see Figure 2):

$$error = (S(\mathbf{x}) - t(\mathbf{x}))^2.$$

Since multilayer perceptrons using neurons with sigmoid output func-tions compute a continuous (monotonic, differentiable) function, the gener-alization is achieved by interpolation using the learning examples as fixed points. In order to optimize generalization, we have to minimize the neural network while keeping the error low. There are two possibilities:

1. minimizing the number of free parameters, that is, the number of neu-rons and weights (see pruning techniques like magnitude pruning, opti-mal brain damage and optimal brain surgeon);

2. minimizing the size of the weights for a given topology (see weight de-cay, weight elimination).

These techniques improve generalization significantly because networks that are too large generalize badly. On the other hand, practical evaluations show that often a network slightly larger than the smallest network with sufficiently low error provides the best generalization.

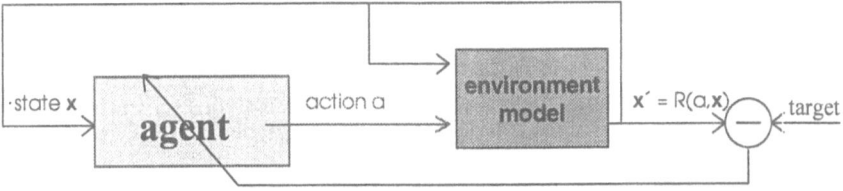

Figure 3: Learning cycle for goal-directed learning with a goal for one step using gradient descent (backpropagation).

Instead of gradient descent, we could also use any of the other techniques for solving the strategy learning problem or for optimizing the agent (see Section 4). If the strategy function of the agent is not differentiable, as in the case of THRESHOLD circuits with binary output functions, we can use hillclimbing or evolution. However, the use of gradient descent (alone or in combination with evolution (Braun 1994, 1997) proved to be significantly more efficient in many practical evaluations whenever the strategy function is differentiable. Additionally, we can improve the performance of the agent by applying the technique of divide-and-conquer in order to build a committee of experts where each expert is specially trained for a subset of the domain.

6.2   Goal-Directed Learning

The problem of goal directed learning is harder than supervised learning because the agent has to anticipate the consequence of the selected action. He gets the goal or target T from the teacher and has to find the action or the sequence of actions which lead to that target.

In the simplest case, one action suffices for achieving the goal. This task is also called *inverse modeling* because a model has to compute the consequence given the action whereas here the learner is given the desired consequence and he has to compute the appropriate action (see Figure 3).

To use gradient descent, we need a differentiable model of the reaction R of the environment in order to "backpropagate" the derivative of the error through this model to the agent (using the chain rule for derivatives). This approach is favorable even if the strategy function is not differentiable, because the gradient of the environment model gives us the direction in which the action should change in order to decrease the error.

If we use a multilayer perceptron both for modeling the reaction of the environment and the strategy function of the agent, we could use backpropagation for optimizing the agent while keeping the weights in the environment model fixed during learning.

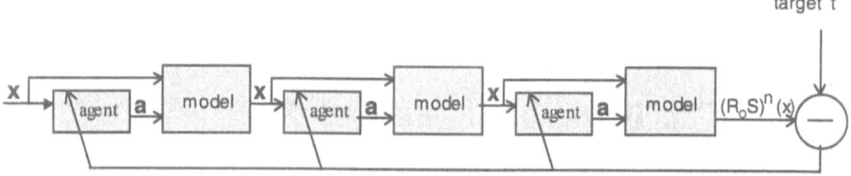

Figure 4: Learning cycle for goal-directed learning with a goal reachable through a decision sequence using the method of unfolding (backpropagation through time).

Significantly harder is the case of the sequential decision problem, that is, when we need a sequence of actions to achieve the goal. In the slang of programming languages we can denote the loop of the sequential decisions as follows:

$$\text{while not } C \text{ do } \mathbf{x} := R\left(S\left(\mathbf{x}\right), \mathbf{x}\right),$$

where C is a stop criterion for terminating the loop (e.g., the goal has been reached). The error of each sequence is the square of the difference between the target and the final state, $\mathbf{x}_f$, at termination:

$$error = \left(\mathbf{x}_f - T\right)^2.$$

To use again gradient descent, we have to unfold the while loop: A while loop is just the multiple sequential execution of the body of the while loop, where the number of executions n is determined by the test condition for the while loop:

$$\mathbf{x}_f := \left(R_M \circ S\right)^n \left(\mathbf{x}\right).$$

In other words, the error is given by the function

$$error = \left(\left(R_M \circ S^n\right)\left(x\right) - T\right)^2.$$

By this unfolding technique we again can use standard gradient descent whenever the reaction model R_M of the environment and the strategy S of the agent are differentiable. This is called *backpropagation through time* for the multilayer perceptron approach (see Figure 4).

A severe problem of this approach lies in the fact that small errors in the environment model may be multiplied by the multiple iteration through unfolding. Nevertheless, Ngyen and Widrow (1990) used this method to solve the difficult problem of backing a truck with one or two trailers up to a dock. Obviously, this task is hard even for human experts, i.e., professional truck drivers.

Another approach is to consider the while loop as a black box with free parameters in the agent (weights). These parameters can be optimized using hillclimbing or evolution but not gradient descent. This approach may be preferable to gradient descent even if the strategy function of the agent and the reaction function of the model is differentiable because of the problems mentioned above.

Moreover, we can reduce the problem of goal-directed learning to reinforcement learning: We assign a positive reinforcement for achieving the goal and a negative reinforcement for the other states. Therefore, we can also use any method for the more general task of reinforcement learning, such as temporal difference learning or dynamic programming (see Section 6.3).

6.3   *Reinforcement Learning*

The scenario of reinforcement learning is the most general, but also the hardest learning problem. The environment (or the teacher) gives positive and negative reinforcement signals (rewards and penalties) depending on the actual state x, and the learner has to optimize its behavior such that the sum of these future reinforcement signals is maximized. For simplicity, we suppose that the domain of possible actions is discrete and finite. Then the task for the agent is to select at each step the action which has the best future reinforcement expectation (see Figure 5). Of course, the future reinforcement depends on the strategy of the agent. Therefore, the agent has to estimate the future reinforcement under the assumption that it keeps its actual strategy fixed. Keeping its strategy fixed, it can optimize or learn the estimation by comparing the actual reinforcement with each of its former estimations. If the agent has learned a sufficiently precise estimation, it can improve its

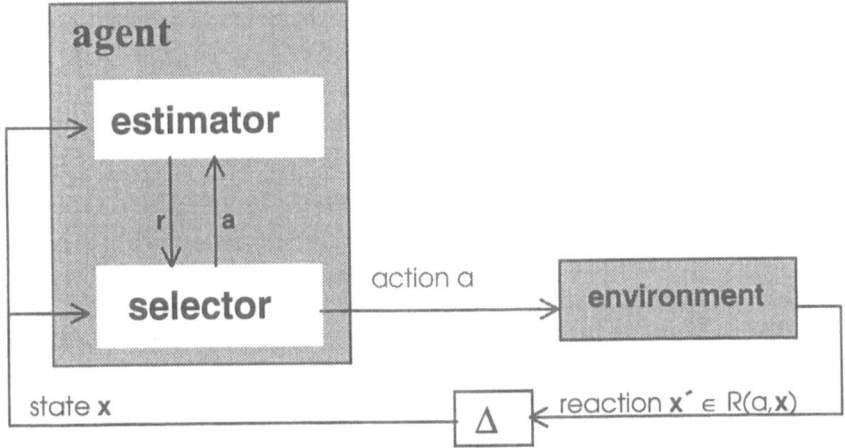

Figure 5: The approach of reinforcement learning uses an estimator for calculating the future reinforcement r. The agent selects the action with the best future reinforcement.

strategy by selecting in each state the action with the best estimated reward according to the old strategy. A simple consideration proves that this new strategy has to be better than the old one unless the old strategy was already optimal. For this new strategy, the agent can optimize again (or learn) the corresponding estimation, and so on. By this reciprocal bootstrapping the agent is guaranteed finally to achieve an optimal strategy.

Therefore, learning a strategy by the agent can be reduced to the problem of learning an estimation of future reinforcements. This approach is well-known to most readers from their experience in game playing. In board games, we select the move leading to the board position with the best estimated reward. The reinforcement signals are winning and losing the game. In games like chess, winning or losing individual pieces provide additional reinforcement signals.

The challenging problem is to find a good estimation (intuition, feeling) for board positions. If an expert with a sufficiently good strategy is available, we could try to imitate this expert in two ways:

1. If the expert also uses a reinforcement estimation function, we could ask him to provide a learning set of prototypical states and the corresponding estimations. Using a multilayer perceptron or other equivalent methods and standard training algorithms such as backpropagation, we could generalize to an estimation function.

2. If the estimation function is not available, the expert could identify the preferred action for pairs of alternatives, and, thus, the preferred successor state. Given such a set of relative ratings, the task is to find an estimation function which is consistent with these ratings. For each data pair the error of an estimation function is defined as the difference between the two corresponding estimations whenever the estimation of the lower rated state is erroneously higher, otherwise the error is 0. Using such a training set of relative ratings, we could also generalize an estimation function just by minimizing these errors for the relative ratings (see Figure 6).

The second approach was successfully applied to training an estimation function for the game of Nine Men's Morris: The achieved performance surpassed the comparable commercial game playing programs (Braun 1991).

Alternatively, the computation of an optimal strategy can be done by dynamic programming: using the Bellman principle compute at step k for each state x, the expected reinforcement r_k of the next k steps for the cheapest path, i.e., for optimally selected actions. Given such an optimal estimator $r = \lim_k r_k$ of the expected future reinforcement, the optimal strategy is just a hillclimbing strategy using this estimator function: Always select the

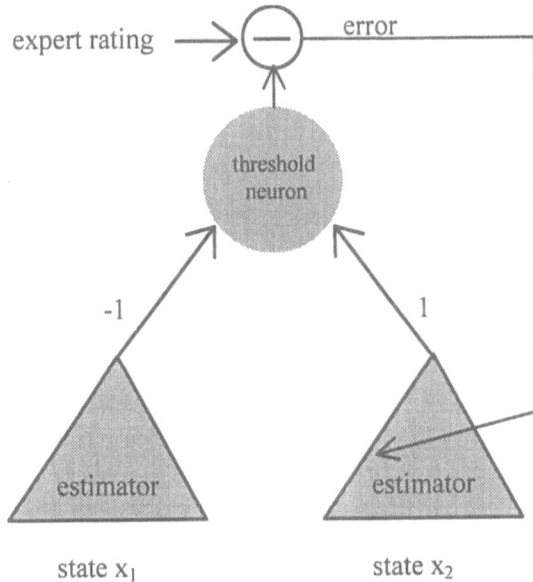

expert rating

error

threshold neuron

-1   1

estimator   estimator

state x_1 state x_2

Figure 6: Training an estimation function using relative ratings (state x_1 is better than state x_2 or vice versa) provided by an expert (both estimator networks are identical, cf. backpropagation with weight sharing).

action by which we reach a state with the best reinforcement estimate of all possible neighbors (neighbors are states reachable by some action). In other words, the selector strategy is to climb to the local optimum of the estimator function despite stochastic perturbations.

Unfortunately, for natural scenarios the set of possible states is far too large to be handled by dynamic programming. Instead, we must generalize the reinforcement estimation function from a sparse set of visited states (the learning set). For example, a mouse which luckily escapes from a cat should generalize this negative reinforcement for the next cat. Temporal difference learning uses a multilayer perceptron for realizing the estimator and optimizes this neural network by minimizing the error using the Bellmann principle: the difference between the actual estimation plus the actual reinforcement and the estimation one step before (Sutton 1988). This approach can also be used for a stochastic scenario.

As already mentioned in Section 4.7, temporal difference learning and other reinforcement learning algorithms can achieve impressive results even in difficult applications, as shown by nearly perfect backgammon playing (Tesauro 1992, 1995), robust broom stick balancing with up to two sticks (one upon another) (Riedmiller 1996) and other fascinating examples (Sutton 1996).

Moreover, it is worth noting that the multilayer perceptron trained by supervised learning (Tesauro 1989) for the game of backgammon was already able to win the computer Olympiad but this winning network was impressively outperformed by a network trained with temporal difference learning (Tesauro 1992, 1995).

7. CONCLUDING REMARKS

For solving sequential decision problems, a human agent typically proceeds as follows. On the first level, the person gets single examples and memorizes them in order to not make the same error twice. The learner may generalize from these examples using a simple Euclidean distance measure. On the second level, the learner tries to generalize by making rules consistent with the encountered examples in order to compress the plain factual knowledge. The learner could also boost its knowledge by calculating the consequences some steps ahead (e.g., a depth-search in game playing). Finally, on the third level, the human uses intuitions or feelings without time-consuming reflection. It seems obvious that we use rational intelligence only for the second level. For the first level, a general purpose memory suffices, that is, in the terms of neural networks, models like learning vector quantization, self-organizing feature maps, or adaptive resonance theory which use a universal distance measure (e.g., Euclidean) for generalization. Only if the task is important enough do we construct a fast, special purpose hardware, that is, neural networks trained as estimators (realizing intuition or feeling) using some reinforcement learning algorithms.

Summarizing, I suggest that by using rational intelligence human agents can close the gap between general purpose memory circuits and special purpose estimator circuits. Obviously, special purpose circuits are expensive to build (hardware costs for neural circuits and software costs and time for training) but fast, whereas rational thinking is slow but generally applicable and therefore "cheap". Therefore, we may conclude that rational thinking is a great tool for solving less important tasks with low development costs (see rapid prototyping) but prerational intelligence achieves better performance.

Systeme, Anwendungen und Produkte der Datenverarbeitung (SAP),
Walldorf/Baden, Germany

REFERENCES

Allender, E. (1989). A note on the power of threshold circuits. *30th Annual Symposium on Foundation of Computer Science* (pp. 580–584). Los Alamitos, CA: IEEE Computer Society Press.

Anderson, J.A. (1977). Neural models with cognitive implications. In S. LaBerge & S.J. Samuels (eds.), *Basic processes in reading perception and comprehension* (pp. 27–90). Hillsdale, NJ: Lawrence Erlbaum Assoc.

Braun, H. (1991). On solving traveling salesman problems by genetic algorithms. *Proceedings of the International Conference on Parallel Problem Solving from Nature PPSN 91* (pp. 128–132). Springer Lecture Notes in Computer Science **496** Heidelberg-Berlin: Springer Verlag.

Braun, H. (1997). *Neuronale Netze – Optimierung durch Lernen und Evolution.* Heidelberg-Berlin: Springer Verlag.

Braun, H. (1999). Evolution – A paradigm for constructing intelligent agents. In H. Ritter, H. Cruse, & J. Dean (eds.), *Prerational intelligence: Adaptive behavior and intelligent systems without symbols and logic,* Vol. 2 (pp. 279–298). Dordrecht, The Netherlands: Kluwer Academic Publishers.

Braun, H., & T. Müller (1990). Enhancing Marr's cooperative algorithm. *Proceedings of The International Neural Network Conference* (pp. 38–41). Paris: Kluwer Academic Publishers.

Braun, H., J. Feulner, & V. Ullrich (1991). Learning strategies for solving the problem of planning using backpropagation. *Proceedings of NEURO-Nimes 91, 4th International Conference on Neural Networks and their Applications.* Nanterre, France: EC2.

Braun, H., & P. Zagorski (1994). ENZO-M – A hybrid approach for optimizing neural networks by evolution and learning parallel problem solving from nature. *Proceedings of the International Conference on Evolutionary Computation* PPSN III.

Chandra, A.K., L.J. Stockmeyer, & U. Vishkin (1984). Constant depth reducibility. *SIAM Journal on Computing* **13**(2), 423–439.

Church, A. (1941) T*he calculi of lambda-conversion.* Princeton, NJ: Princeton University Press.

Fahlman, S., & C. Lebiere (1990). *T*he cascade correlation learning architecture. *Technical Report CMU-CS-90-100*, Pittsburgh: PA: Carnegie Mellon University.

Fogelman, F., E. Goles, & G. Weisbuch (1983). Transient length in sequential iterations of threshold functions. *Discrete Applied Mathematics* **6**, 95–98.

Furst, M., J.B. Saxe, & M. Sipser (1984). Parity, circuits and the polynomial time hierarchy. *Mathematical Systems Theory* **17**(1), 13–27.

Goldmann, M., J. Hastad, & A. Razborov (1992). Majority gates vs. general weighted threshold gates. *Proceedings of the 7th Annual Structure in Complexity Theory Conference* (pp. 2–13). New York: IEEE Computer Society Press.

Goles,E., & J. Olivos (1981). The convergence of symmetric threshold automata. *Information and Control* **51**, 98–104.

Gödel, K. (1931). Über formal unentscheidbare Sätze der *Principia Mathematica* und verwandter Systeme, I. *Monatsschrift Mathematische Physik* **38**, 173–1981. (English translation: *On formally undecidable propositions of* principia mathematica *and related systems* (translated by B. Meltzer). New York: Basic Books.)

Grötschel, M. & O. Holland (1991). Solution of large-scale symmetric traveling salesman problems. *Mathematical Programming* **51**, 141–202.

Grossberg, S. (1987). *The adaptive brain I/II.* Amsterdam: Elsevier.

Hajnal, A., W. Maass, P. Pudlák, M. Szegedy, & G. Turán (1987, October) Threshold circuits of bounded depth. In *28th Annual Symposium on Foundations of Computer Science* (pp. 99–110). New York: IEEE Computer Society Press.

Haken, A., & M. Luby (1988). Steepest descent can take exponential time for symmetric connection networks. *Complex Systems* **2**, 191–196.

Haken, A. (1989). *Connectionist networks that need exponential time to stabilize.* Unpublished manuscript. Deptartment of Computer Science, University of Toronto.

Hartley, R., & H. Szu (1987). A comparison of the computational power of neural networks. *Proceedings of the 1987 International Conference on Neural Networks*, Vol. 3 (pp. 15–22). New York: IEEE Computer Society Press..

Hassibi, B. & D.G. Storck (1993). Second order derivatives for network pruning: Optimal brain surgeon. In S.J. Hansen, J.D. Cowan, & C.L. Giles (eds.), *Advances in neural information processing systems 5* (NIPS 5) (pp. 164–171). San Mateo, CA: Morgan Kaufmann.

Hastad, J. (1992). *On the size of weights for threshold gates.* Unpublished Manuscript.

Hong, J. (1986). *Computation: Computability, similarity and duality.* London: Pitman Publishing.

Hong, J. (1987). *On connectionist models.* Technical Report 87-012, Department of Computer Science, University of Chicago.

Hopfield, J.J. (1982, April). Neural networks and physical systems with emergent collective computational abilities. *Proceedings of the National Academy of Sciences, USA,* **79** (pp. 2554–2558).

Hopfield, J.J., & D.W. Tank (1985)."Neural" computation of decisions in optimization problems. *Biological Cybernetics* **52**, 141–152.

Jordan, M., & R. Jacobs (1994). Hierarchical mixtures of experts and the EM algorithm. *Neural Computation* **6**, 181–214.

Karmarkar, N. (1984). A new polynomial-time algorithm for linear programming. *Combinatorica* **4**, 373–395.

Kleene, S.C. (1956). Representation of events in nerve nets and finite automata. In C.E. Shannon & J. McCarthy (eds.), *Automata Studies. Annals of Mathematics Studies* 34 (pp. 3–41). Princeton, NJ: Princeton University Press.

Kohonen, T. (1982). Self-organized formation of topologically correct feature maps. *Biological Cybernetics* **43**, 59–69.

Littmann, E., & H. Ritter (2000). Modularization by cascading neural networks. In H. Ritter, H. Cruse, & J. Dean (eds.), *Prerational intelligence: Adaptive Behavior and intelligent systems without symbols and logic, Vol. 2* (pp.165–181). Dordrecht: Kluwer Academic Publishers.

Maass, W., G. Schnitger, & E. Sontag (1991). On the computational power of sigmoid versus Boolean threshold circuits. *Proceedings of the 32th Annual IEEE Symposium on Foundations of Computer Science* (pp. 767–776). New York: IEEE Press.

McCulloch, W.S., & W. Pitts (1943). A logical calculus of ideas immanent in nervous activity. *Bulletin of Mathematical Biophysics* **5**, 115–133.

Mealey, G.H. (1955). Method for synthesizing sequential circuits. *Bell Labs Technical Journal* **34**, 1045–1079.

Moore, E.F. (1956). Gedanken-experiments on sequential machines. In C.E. Shannon & J. McCarthy (eds.), *Automata Studies. Annals of Mathematics Studies 34* (pp. 129–153). Princeton, NJ: Princeton University Press.

Muroga, S., I. Toda, & S. Takasu (1961, May). Theory of majority decision elements. *Journal of the Franklin Institute* **271**, 376–418.

Neumann von, J. (1951). The general and the logical theory of automata. In L.A. Jeffress (ed.), *Cerebral mechanisms in behavior: The Hixon symposium* (pp. 1–32). New York: Wiley.

Nguyen, D., & B. Widrow (1990). The truck backer-upper: An example of self-learning in neural networks. In R. Eckmiller (ed.), *Advanced neural computers* (pp. 11–19). Amsterdam: North Holland.

Padberg, M. & G. Rinaldi (1987). Optimization of a 532-city symmetric traveling salesman problem by branch and cut. *Operations Research Letters* **6**, 1–7.

Parberry, I. (1994). *Circuit complexity and neural networks*. Cambridge, MA: MIT Press.

Riedmiller, M. (1996). Learning to control dynamic systems. In R. Trappl (ed.), *Proceedings of the European Meeting on Cybernetics and System Research EMCSR 96*. London: World Scientific.

Ritter, H., T. Martinetz, & K. Schulten (1990). *Neuronale Netze*. Reading, MD: Addison Wesley.

Sigelman, H.T., & E.D. Sontag (1992). On the computational power of neural nets. *Proceedings of the 5th Annual Workshop on Computational Learning Theory* (pp. 440–449). New York: ACM Press.

Sipser, M. (1983). Borel sets and circuit complexity. *Proceedings of the Fifteenth Annual ACM Symposium on Theory of Computing* (pp. 61–69). New York: ACM Press.

Sontag, E.D. (1992). Feedforward nets for interpolation and classification. *Journal of Computational System Science* **45**, 20–48.

Stahlberger, A., & M. Riedmiller (1997). Fast pruning and feature extraction by removing complete units. In M.C. Mozer, M.I. Jordan, & T. Petsche (eds.), *International Conference of Neural Information Processing Systems, NIPS 9* (pp. 655–661). Cambridge, MA: MIT Press.

Sutton, R.S. (1988). Learning to predict by the methods of temporal differences. *Machine Learning* **3**, 9–44.

Sutton, R.S. (1996). Generalization in reinforcement learning: Successful examples using sparse coarse coding. In D.S. Touretzky, M.C. Mozer, & M.E. Hasselmo (eds.), *Advances in Neural Information Processing Systems 8* (pp. 1038–1044). Cambridge: MIT Press.

Tesauro, G.J. (1989). Neurogammon wins computer Olympiad. *Neural Computation* **1**, 321–323.

Tesauro, G.J. (1992). Practical issues in temporal difference learning. *Machine Learning* **8**, 257–277.

Tesauro, G.J. (1995). Temporal difference learning and TD-Gammon. *Communications of the ACM* **38**(3), 58-68.

Turing, A.M. (1936). On computable numbers with an application to the Entscheidungsproblem. *Proceedings of the London Mathematical Society* **2**(42), 230–265.

HELGE RITTER* and HANS-ULRICH BAUER**

MATHEMATICAL PERSPECTIVES ON PRERATIONAL INTELLIGENCE

1. INTRODUCTION

Mathematical laws are rigid and precise; their aim is to constrain and predict. A mathematical description leaves no leeway for interpretation or intuition. In contrast, the study of prerational intelligence confronts us with a host of notions for which we have an intuitive understanding, but which often turn out to be hard to formalize. If this is so, which aspects of intelligence are then amenable to mathematical analysis and understanding? How can we formalize such multi-facetted aspects of prerational intelligence as robustness, flexibility, learning, complexity, goal-seeking, and self-organization, to name just a few. And how does the organization of coordinated activity within a system, be it on the level of a neural network, an entire organism or a society of social insects, arise from the interactions of constituents of the system? What mathematical abstractions capture in a sufficiently parsimonious way the challenging yet volatile aspects of the complex patterns of interaction between an intelligent agent and its environment, which we tend to call intelligent, such that a formal mathematical analysis or, at least, a simulation on a digital computer can be performed? To what extent can mathematical analysis help to build artificial systems that appear to an observer to behave intelligently? Vice versa, can the goal of analysing prerational intelligence in mathematical terms suggest the invention of interesting new notions in the field of mathematics itself? And if the answer to some of these questions is positive, what are the subjects mathematical analysis should focus on?

In this chapter we would like to discuss these issues. It goes without saying that we will leave the majority of these questions without a final answer. What we do attempt, however, is to point out where mathematical tools and theories exist that may contribute to a more precise understanding of issues connected with prerational intelligence, and where known mathematical results can already provide some partial answers to non-trivial questions in the field.

Mathematical models related to prerationally intelligent systems can be formulated on different levels of organization. The possibilities range from the level of small neural networks through larger neural systems up to a level

J. Dean et al. (eds.), Prerational Intelligence: Interdisciplinary Perspectives on the Behavior of Natural and Artificial Systems, 201–245.

where the modeling of the behavior of an entire agent, or even a number of agents in their interaction with the environment, is attempted. Consequently, the complexity of the constituents of the models varies greatly: single neurons on the lowest level, entire subsystems or modules on the intermediate levels, and entire organisms on the highest level. Corresponding to this huge variety, many specific models have been developed to capture in detail aspects of particular systems (e.g., specific neuronal oscillator models to capture the details of a central pattern generator). Even within each level, models can be formulated at different levels of abstraction and with different modeling objectives in mind. One can have a model that is close to the actual physical implementation of a system, with the aim of analysing which features of the physical implementation are responsible for some functional aspect under study. Examples of this detailed level of description are compartment models for individual neurons. One can also entirely ignore the structure of the physical implementation and focus on a number of derived quantities that appear suitable for describing certain properties of the system. An example of this more abstract level is the McCulloch-Pitts description for single neurons. Many biologically plausible models for neural information processing are close to the first kind of approach. Models based on explicit symbol processing, and a number of other approaches from the field of artificial intelligence or psychology fall into the second category. Clearly, a theoretical description of a specific system on one of these levels of organization, following a specific modeling objective on a particular level of abstraction, requires mathematics to a possibly large degree. Yet, a list of all of the different approaches would not only be exceedingly long, it would be barely possible to compile and it would be boring to read. In particular, such a list would obscure commonalities of underlying mathematical principles.

These unifying mathematical principles and approaches are the focus of the present chapter. They correspond to overlapping or general aspects of prerationally intelligent systems. The commonalities of the systems include, for example, structures with a large number of nonlinearly interacting constituents, each of which has a correspondingly small impact, the limited number of discriminable states for each constituent, the lack of dedicated central control structures, or the adaptability of the system under external influences. In the following sections we briefly discuss several mathematical frameworks, which are of similar generality. They can operate on many of the above mentioned levels, and they make commonalities of the underlying systems more transparent.

The second section of this chapter is devoted to coding, that is, to the "appropriate representation of information", a central issue for any intelligent system. A related mathematical paradigm is pattern formation and pattern recognition, which we will discuss in a third section. The fourth section is focused on dynamical systems, a framework for the description of the time evolution of (nonlinear) systems. The range of applicability of dynamical systems theory goes way beyond prerational intelligence; this mathematical abstraction is of universal importance as long as a system is not static. In Section 5, the notion of optimization is discussed. Optimization techniques are broadly applicable in many fields of science, including several domains which are related to prerational intelligence. Finally, in Section 6, we focus on a theoretical field which is central to intelligent systems, learning theory.

2. CODING

In some way or other, intelligent behavior must reflect certain aspects of the environment. If we want to understand this behaviour as the result of some computational process, we need to analyse how this process operates. In particular, we need to know what information about the environment is utilized and is, consequently, encoded, how it is encoded, and how it is transformed by the computational process.

In biological systems, neural activity patterns are the basis of any such process. They somehow "represent" one part of the information that is processed and transformed in the brain. Another part of this information is hidden in the structure of the "hardware": the neural connectivity, the momentary spatial distribution of neurotransmitters and neuromodulators, and probably additional factors. It is one of the major goals of brain theory to understand how information is encoded in these different levels and how this encoding enables the necessary computations to be achieved by the parallel computation of many (relatively) simple constituents, using local rules.

A first question is: how much information can we expect to be represented in a single neuron? An estimate can be found by considering the activity of a single neuron as a source of information about some stimulus, or for some discrimination task in which the neuron is assumed to participate. One then can apply information theory to compute the information content of the neural response with regard to the presented stimuli. This method has led to estimates of a few bits of information coded in individual neuronal signals (Optican & Richmond 1987; Bialek et al. 1991; Richmond & Optican 1990). A next question is how this information content is affected when a signal is successively transformed by a chain of neural centers (Barlow 1972). Studies of this issue for the visual and for the taste modalities indicate that the information about stimulus identity increases towards the higher processing stations (Rolls 1989).

In a related approach the question is slightly reformulated: how much of the stimulus can be reconstructed from the neural activity pattern? This would somehow require inversion of the encoding process that took place in the afferent neural pathways of the neuron under study. Artificial neural networks have provided a means to approach this question: they can be trained to re-transform the neural activity into the identity or more elaborate features of the original stimulus and the amount of correlation achievable in this way gives a clue to how much and what kind of information about a stimulus is encoded in the neural signal. This work, too, has confirmed the estimate of a few bits of information per neuron (Bialek 1992). For simple (e.g., binary) discrimination tasks, this may be sufficient to allow a reliable decision on the basis of the activities of single neurons. In fact it has been shown that the discrimination reliability achievable on the basis of spike signals of individual neurons in the visual cortex (area MT in monkeys) equals or even (slightly) exceeds the reliability expressed in the behaviour of the entire animal during a psychophysical experiment on the same task (Newsome et al. 1989). Exploiting spike trains that were recorded from an array of electrodes under different stimulus conditions (different orientations of a moving bar) it has been possible to decode the spike train patterns and to classify the stimulus conditions to a quite high precision (Krüger & Beckers 1991; Radons et al. 1994). In other investigations, behavioral responses of a monkey were found to be correlated with particular states observed in multielectrode spike trains (Gat & Tishby 1993). An open, but much debated question in this regard is whether synchronized, oscillatory discharge patterns can be detected by subsequent processing stages in a background of uncorrelated, noisy discharges: this is the readout problem. Even though the coincidence detection properties of neurons make such a coding scheme attractive, the high variability observed in recordings of synchronized neuronal activity (see, e.g., Kreiter & Singer 1992) might spoil this concept.

The amount of information in a signal by itself is a rather coarse characteristic of neuronal responses. More insights can be gained by considering specific properties of codes. These can be cast into a mathematical formulation, from which, in turn, predictions can be derived. This strategy has successfully been applied, for example, to the analysis of receptive field properties. In the following, we list some of the major issues that have been considered so far. A more elaborate discussion of many of these points can be found in Bialek (1992).

Precision.   The amount of information in neuronal signals depends on the precision of their transmission. The usefulness of this information for subsequent computations depends also on the robustness of signal processing in neural systems with respect to noise, tolerances and failures. These issues have been extensively studied in artificial neural networks, where one can introduce various kinds of failures and study their impact on processing properties, such as retrieval accuracy and memory capacity in associative memories, or classification rate in neural classifiers (see, e.g., Amit 1989; Debenham & Garth 1989, see also the previous chapter).

Redundancy.   Good use of resources implies that the encoding should have little redundancy. A first step to reduce redundancy is decorrelation. Codes in which the individual elements are uncorrelated also facilitate the detection of newly appearing associations (i.e., of associations appearing after the code has been constructed). Local decorrelation leads to feature detectors that can be described as spatial convolution filters, using orthogonal function systems. Assuming a suitable probability density for the stimuli, Hebb-type learning can lead to the formation of such individual filters (Oja 1982). Anti-Hebbian adaptation rules for connections between different filter channels can further produce a redundancy-reduced coding of input information in an extended neural structure (Barlow & Földiák 1989; Rubner & Tavan 1989).

Decorrelation is only a first step towards redundancy reduction. More sophisticated reduction schemes have been realized with optimization approaches based on artificial neural networks (Schmidhuber 2000). The significance of the resulting codes with respect to brain theory is still a subject of discussion.

Error-tolerance.   The good use of resources by reduction of redundancy advocated in the previous paragraph is limited by the somewhat opposite desire for error-tolerance. Codes without any redundancy cannot be error-tolerant; compensation for errors requires some amount of redundancy. Which form this redundancy takes depends on the type of errors that must be repaired. The requirements of a particular error correction can be expressed as terms in a cost function. To find the associated code then requires minimization of this cost function. Following this general approach, one can, for example, derive optimal filter characteristics of retinal ganglion cells for varying types of noise and noise levels, which coincide with experimentally observed filter curves (Atick & Redlich 1992). This method of constructing (approximately) optimal codes can be related to learning approaches. In particular, Hebbian learning can be regarded from this viewpoint, as well as the formation of topographic maps (Luttrell 1989).

Coarse coding.   Which coding schemes can achieve precise encoding of values when the available elements are of a low precision? Analysis of this question has led to the concept of coarse coding. A coarse code achieves a high precision for the representation of a signal by suitably combining the responses of a (usually) large number of low-precision elements with broad and overlapping tuning curves. For the case of a superposition of the individual contributions, this has been investigated in mathematical detail (Baldi 1988; Heiligenberg 2000). Coarse coding is intimately related to the phenomenon of hyperacuity, which denotes a perceptual discrimination precision that exceeds the precision of the individual receptors (see, e.g., Westheimer 1977; for a discussion of hyperacuity in the context of pre-rational intelligence, see Wehrhahn 2000). Coarse codes also turn out to be one way to meet the diverging objectives of precision, redundancy and error-tolerance.

Invariance.   Many computations have the goal of combining high discrimination with respect to a small set of stimulus properties and invariance with respect to most others. The study of invariance properties can thus help find codes that are suitable for certain tasks. Spatial filters can be classified with the help of differential geometry according to the types of local transformations of the view field to which they are sensitive. This permits a classification of the receptive field properties that are useful for detecting edges, curvature of lines or of surfaces, or three-dimensional rigid body motions (Koenderink 1989).

Representation vs. computation.   The way some information is encoded has a strong influence on the effort required for using it for a particular task. A simple example is the representation of numbers. If the task is to check for divisibility by, say, 17, a number representation in that base is superior to the usual decimal representation. The reason is that this choice makes the property of interest much more "explicit", that is, retrievable by a short and simple computation, as compared to a decimal representation. What can theory tell us about properties of codes that make them particularly well-suited for parallel computation with many neuron-like elements? In the realm of associative memory, for example, sparse coding schemes have been shown to allow an increase in the number of patterns that can be stored (Gardner 1987; Tsodyks & Feigel'man 1988). However, at the same time, the information content per pattern decreases, leaving the total amount of stored information approximately constant. However, these sparse codes seem more in accord with biological reality and also have better superposition properties (Atick & Redlich 1990).

Relations. It is often necessary to code not just information about individual object properties, as in the response of a single, feature-selective cell, but also about their mutual relations. The coding of such relations in neural networks is still a controversial matter, in particular, if several such relations have to be coded simultaneously. One proposal (Milner 1974; v.d. Malsburg 1981, 1986), which has been more closely investigated mathematically (see, e.g, Sompolinsky et al. 1991), is the use of temporal synchronization among oscillating elements.

3. PATTERN FORMATION AND PATTERN RECOGNITION

In the previous section we pointed out properties of (neural) codes which affect the emergence of prerational intelligence. Now we turn to the more general notion of patterns. Patterns are also intimately connected with many aspects of prerational intelligence. Intelligent behavior can be viewed as the establishment of particular kinds of structured spatio-temporal patterns of interaction with the environment. On the sensory side, this may require the ability to recognize certain types of patterns; on the motor side, certain temporal patterns or rhythms must be generated. Both processes are usually linked in a complex fashion. Finally, the neural substrate itself is a structure in which we find patterns, for example, of connectivity, on various levels of organization.

Let us first consider how a "pattern" can be characterized in a more formal way. A typical property of a pattern is that it exhibits a certain amount of redundancy. Exploiting this, one can use a given part of a pattern to make predictions about other parts. An simple, yet powerful type of such a redundancy is a symmetry. In mathematics symmetries have been extensively used to describe and classify many types of patterns. A famous example is the classification of the possible crystal shapes into the 32 point groups of crystallography. Another example from contemporary mathematics is the exhaustive classification of all tilings of the plane into different classes. In the context of prerational intelligence, symmetry can also be found on various levels. On the highest level, we observe the bilateral symmetry of the brain hemispheres. As an example on a lower level, the detection of symmetries plays an important role in vision. Here, the symmetry of objects can be utilized to reduce the number of required 2d views of objects which are necessary to infer their 3d structure (Poggio & Vetter 1992). Psychophysical evidence indicates that symmetry is indeed exploited in this way (Vetter et al. 1994). Recent investigations into the recognition of symmetric vs. asymmetric patterns revealed that a preference for symmetric patterns can result as a by-product of an evolutionary process (Enquist & Arak 1994; Johnstone 1994).

Different, less constraining forms of redundancy are correlations within and between signals $x(t), y(t)$,

$$C_{xx}(\tau) = \langle x(t)x(t+\tau)\rangle_t \qquad \text{(Autocorrelation)} \qquad (1)$$

$$C_{xy}(\tau) = \langle x(t)y(t+\tau)\rangle_t \qquad \text{(Crosscorrelation)} \qquad (2)$$

Usually, nearby elements of a pattern tend to be correlated in their properties. Spatial correlations between two patterns are exploited, for example, to solve the correspondence problem occurring during the detection of motion or depth from pairs of images.

Correlations may be simultaneously exhibited on very different scales. Consider a photograph of a car as an example. Here, the color of nearby pixels is correlated. However, there are also long-range correlations: given the portion of the image that shows the front wheel, one can make a good estimate of where the rear wheel is located in the image. Moreover, one will have a great deal of information about the image at that distant location, since front and rear wheel usually look similar. Long range correlations of this type are valuable constraints that may be used by computer vision systems (Sagerer & Niemann 1997).

The concept of correlation allows a generalization of the notion of symmetry: to test for the presence of a symmetry in a pattern, one considers the correlation between the pattern x and its transform Tx under the symmetry operation T that is under consideration. If the pattern exhibits the symmetry precisely, $Tx = x$ and the correlation will be large. Otherwise, Tx will deviate from x more or less, leading to a smaller correlation that indicates that the symmetry is only approximate or not present at all.

The evaluation of correlations is not the only mathematical technique to pinpoint structure in a data set. In particular, one cannot conclude from a vanishing correlation the absence of a regularity that may qualify as a "pattern". Visually detectable evidence for this statement are texture patterns that coincide in their global second order statistics (and therefore in their cross-correlation), or even in their third order statistics, but that are still discriminable, even preattentively (Julesz 1981). A measure for the presence of regularities which is based on information theory and which goes beyond auto- or cross-correlation is the concept of *mutual information*. If x and y denote two attributes of a pattern (such as the color at two different locations in an image), we may consider them as random variables with a joint probability density $p(x, y)$. If knowledge of x never gives any clue about the value of y, the two random variables must be statistically independent and $p(x, y)$ will be the product $p(x)p(y)$ of the individual probability densities. Otherwise, knowledge of one value provides some non-vanishing information

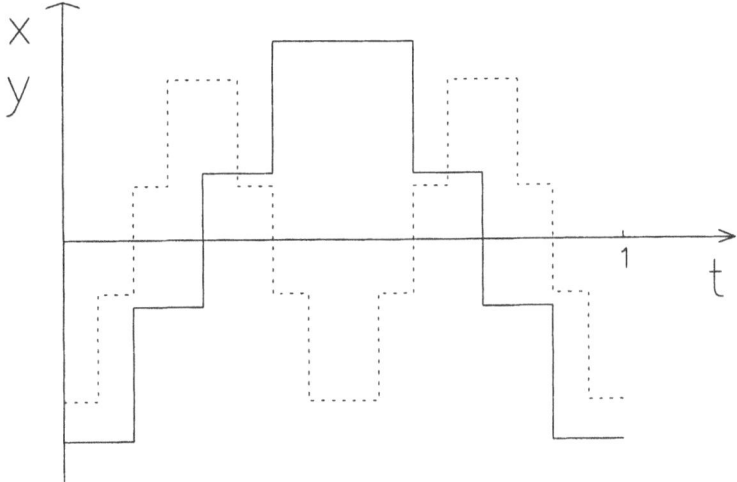

Figure 1: Two signal trains $x(t)$ (solid) and $y(t)$ (dashed), with four different, equiprobable values (states) for x, y, respectively, and eight equiprobable combination states for (x, y). Clearly, we have $<x(t)>_t = 0$, $<y(t)>_t = 0$, and for symmetry $<x(t)y(t)>_t = 0$. Under the assumption of periodic signals, $x(t + 1) = x(t)$, $y(t + 1) = y(t)$, delayed crosscorrelations vanish as well, $<x(t)y(t+\tau)>_t = 0$, again for symmetry reasons. The information (in bits) conveyed by an observation of x amounts to $I_x = -\sum_x p(x)\mathrm{ld}p(x) = 2$ bits, analogously $I_y = 2$ bits. The mutual information between x and y yields (in bits) $M = \sum_{x,y} p(x,y)\mathrm{ld}p(x,y) + 4$ bits $= 1$ bit, that is, we know from the measure of one signal 1 bit about the value of the other signal. For example, if we measure for x the most positive value (or the most negative), then we know without further measurement, that y will be negative (1 bit of information). The same outcome results (vanishing crosscorrelation, but nonvanishing mutual information), when we do not observe x, y directly, but, say, spike trains with firing probabilities which are derived from x and y.

about the possible value of the other. The expected amount of information is given by the mutual information

$$M(x,y) = \sum_{x,y} p(x,y) \ln p(x,y) - \sum_x p(x) \ln p(x) - \sum_y p(y) \ln p(y)$$

(3)

In general, a vanishing mutual information implies zero correlation. The converse, however, is not true, except if $p(x, y)$ is Gaussian. In this case mutual information and correlation are related in a simple way. A simple example for two signals, which have a vanishing cross-correlation and a nonvanishing mutual information, is provided in Fig. 1.

These concepts or advanced methods derived from them have to be used when we want to detect the presence of patterns of unknown characteristics in given signals, or when we wish to quantify properties of the signals. Signal in the present context of prerational intelligence means neural spike trains, local field potentials, EEG-signals or the spatio-temporal activity distributions that have become measurable with modern imaging techniques in various parts of the brain. One example for an advanced method is the time-resolved mutual information which can keep track of the possibly non-stationary degree of dependence between two dynamical signals. It is derived from the usual mutual information by using a particular summation procedure to evaluate the individual terms in Eq. (3). The modified summation expresses the contributions to the overall mutual information M as a sum over time instead of a sum over states. With the time-resolved version of the mutual information, a time-dependent coupling between two local field potential signals from cat visual cortex has been identified (Pawelzik et al. 1993).

Let us now proceed from a general evaluation of correlations to the more specific domain of pattern recognition. The evaluation of particular correlations forms the basis of many traditional approaches to the analysis of patterns. A typical feature for many of these is the convolution of the pattern with a number of filter kernels, which can also be regarded as a computation of the correlations of the pattern with the filter kernels. Choosing sinusoids as filter kernels, for example, leads to Fourier analysis. Via the choice of the filter kernels one can determine which features of the pattern are to be made explicit in the computed correlation coefficients. In the case of Fourier analysis, these features are spatial or temporal frequencies, at the expense of any information about location. Making other features explicit leads to different filter sets. The requirement of an "optimal" compromise between spatial and frequency resolution leads to Gabor filters (Gabor 1946)

$$s(t) = e^{(-(t-t_0)^2/\sigma^2 + i\omega t)},$$

a sine wave with a Gaussian envelope. In two dimensions these arguments lead to optimal filter properties which resemble those of simple cells in cat visual cortex (Daugman 1985).

In the neurobiological realm, filter or pattern recognition properties of single neurons have been extensively investigated, mainly in the sensory domains, in particular in the primary visual cortex. A major paradigm for these studies is that of the "feature detector": a neuron that responds to a certain range of features, such as light bars within a certain range of orientations. Even though the characterization of feature detection properties might seem at first glance to be an experimental issue, a few mathematical remarks can be made. We already dwelt on the advantages of non-redundant codes in the section on coding. The idea of maximizing the output variance of a neuronal projection chain in response to "pseudovisual" stimuli has helped to understand the self-organization of successive receptive field properties of individual cells in a layered system which is analogous to early visual processing (Linsker 1986).

The assumption that the responses of visual neurons are such that they minimize a suitable cost function that is based on the correlation or the mutual information of their output activities is an alternative approach to make predictions about the nature of the features to which these neurons should respond. Maximizing the correlation or the mutual information between otherwise independent pattern recognizers has led to new unsupervised learning schemes for pattern classification that provide working examples of mechanisms by which feature integration might be achieved within a modular system (Becker & Hinton 1992; de Sa 2000).

Requiring local correlations for the properties of feature detectors in a two-dimensional neural sheet together with a minimal amount of diversity in the response properties of the detectors enforces particular patterns for the spatial arrangement of the feature detectors that closely resemble many aspects of the topographic maps in the visual cortex. If applied on a more abstract level, the same principles may lead to spatial maps of semantic information (Ritter & Kohonen 1989).

The last example already leads into the complementary field of pattern formation. On the level of the neural substrate, one would like to find principles that underlie the formation of the observed neural connectivity patterns. Topographic maps are perhaps the most conspicuous patterns of this kind. Mathematical models for their formation can be based on rather simple mechanisms: short range lateral excitation, resulting in a local spread of activation, and long range lateral inhibition, resulting in competition between neurons (v.d. Malsburg 1973; Willshaw & v.d. Malsburg 1976). The perhaps simplest model for map formation processes has been put forward by Kohonen (1982, 1989). It characterizes each feature detector by its position r in a two-dimensional sheet and by a parameter vector w_r. The model assumes that randomly incoming stimuli x cause adaptive changes that are confined to a neighborhood of the best-responding feature detector s (the location s being defined by maximizing the dot product $w_r \cdot x$ with respect to r). This is mathematically expressed by the adaptation law

$$\Delta w = h(\|r - s\|)(x - w) \qquad (4)$$

where $h(\cdot)$ is a bell-shaped function, centered at the origin. It ensures that any changes due to Eq. (4) are confined to the immediate vicinity of the feature detector at position s.

The simple-looking adaptation law (4) is able to transform the random sequence of "stimuli" x into rather complex spatial patterns of the parameters w. The evolved structures result from the tendency of (4) to form a dimension-reduced mapping of the space of the stimuli x that also reflects the non-uniform sampling from the stimulus probability density. As already indicated, suitable stimulus probability densities can lead to surprisingly good modelings of topographic maps in the visual cortex (Obermayer 1990), a property which was the original motivation for the invention of the process and its study.

However, here we have a good example how the study of a particular self-organization process, the formation of topographic maps, can lead to a mathematical model that then turns out to capture a principle of considerably wider generality. It turns out that the same process can be used for solving tasks that seem at first sight quite unrelated to its original motivation.

These more general tasks and problems include more than just a variety of map formation phenomena in various modalities of the brain (including the visual (Wolf & Bauer 1994), the somatosensory (Ritter & Schulten 1986), or the auditory domain (Martinetz et al. 1988). A much more abstract type of task is to find short paths connecting a given number of points. By itself, or as an archetypical representative of the class of np-hard optimization problems, this "traveling salesman problem" may also have some relevance for simpler organisms and is, therefore, related to prerational intelligence. Since it is a difficult optimization problem, it has elicited the development of numerous algorithms for its solution, many of them complex and requiring rather intricate symbol processing.

Surprisingly, the simple-looking process (4) is able to generate rather good solutions for the TSP-problem. All that is required is to arrange the "feature detectors" on a closed ring (instead of a two-dimensional sheet), and to concentrate the stimulus density at the locations of the points to be visited. Then, starting from an initially circular path, the local adaptations specified by (4) will gradually deform the ring into a closed path that visits all points and the length of which – while usually not attaining the minimum – is very short. This property of map self-organization is not restricted to the specific form of Eq. (4), but occurs as well in related algorithms (like the elastic net algorithm (Durbin & Willshaw 1987)).

There are many other interesting ways for generating complex patterns from simple rules that may provide us with interesting paradigms for the emergence of order and structure in systems of many simple constituents.

One branch of pattern formation mechanisms is based on reaction-diffusion equations. In the most simple case, there is only a single equation describing the diffusion and the decay of some either real or fictitious substance. If we denote the concentration of this substance at a location r and time t by $u(r, t)$, the simplest hypothesis is a diffusion equation with some decay term:

$$\partial_t u(r,t) = c\Delta u(r,t) - \gamma u(r,t) + h(r) \tag{5}$$

Here, $h(r)$ describes the presence, of some source, where new substance is injected (or absorbed, if $h(r) < 0$). By suitable choices for $h(r)$ and the boundary conditions (where the propagation of the substance may be blocked, or where substance may be absorbed) the diffusion process (5) can be viewed as the equivalent of the parallel exploration of some "terrain" by an infinite number of continuously distributed agents. The stationary solution $u(r, t = \infty)$, therefore, encodes information about the position of the sources $h(r)$, as well as of any obstacles that have blocked the diffusion of the substance u. In particular, starting at some initial position s and following the gradient ∇u will generate a path that avoids all obstacles and that leads to one of the sources represented by $h(r)$. Again, we have a simple dynamical process that offers a basis for solving such tasks as obstacle avoidance, path planning and robot navigation. Of course, the simple approach (5) still has several limitations that motivate a number of non-linear extensions to improve performance.

With regard to pattern formation, the process (5) is still rather limited. A significant next step is to subject the hitherto static sources $h(r)$ also to a dynamical law.

This can be achieved by stipulating an autocatalytic process for the fictitious substance u. A simple way to implement that is to make the source term $h(r)$ proportional to some power of u. However, this destabilizes the system unless some source of inhibition is introduced that can balance the autocatalysis of u. Following work by Gierer and Meinhard (1972) which is based on an earlier proposal by Turing (Turing 1952) this can be achieved by introducing a second diffusing substance, an "inhibitor", described by a concentration $v(r,t)$ and subject also to a diffusion law, but with a different (larger) diffusion constant. The system of coupled differential equations

$$\partial_t u(r,t) = c_u \Delta u(r,t) - \gamma_u u(r,t) + h_0 + u(r,t)^2/v(r,t) \tag{6}$$
$$\partial_t v(r,t) = c_v \Delta v(r,t) - \gamma_v v(r,t) + u(r,t)^2 + h_1 \tag{7}$$

is capable of forming very complex spatial structures, depending on the values of the parameters and of the boundary conditions. The generated structures and the dynamics of their formation can model many aspects of morphogenetic processes, such as the development of vessels in a leaf, the arrangement of leaves and buds around a stem, and many more.

Other forms of reaction-diffusion equations describe the interaction of a diffusing substance with an array of periodic clocks. The clock models an oscillating chemical reaction that periodically creates a pulse of some substance that in turn affects the phase of the oscillator, but that also is subject to diffusion, in this way coupling the phases of nearby oscillators in a non-linear fashion. Reaction-diffusion equations of this type have be found suitable to describe pattern formation processes underlying such seemingly different phenomena as the generation of the heart rhythm and the aggregation and development of the slime-mold *Dictiostelium discoideum* (Winfree 1987).

The actual numerical integration of the continuous differential equations above can be very time consuming on sequential computers. It turns out that many of the pattern formation phenomena that are described by the continuous differential equations can be described qualitatively as well by using strongly simplified "discrete caricatures" in the form of cellular automata with suitably chosen rules to mimic the original non-linear dynamics. This brings these pattern formation processes even closer to a format in which many simple constituents interact such that some non-trivial task, here manifested in the formation of complex spatio-temporal structures, is accomplished.

There are a number of interesting resemblances between cellular automata and neural networks. Both types of systems contain large numbers of simple constituents that are characterized by a small number of internal states that are continuously updated on the basis of the states of the constituents in some neighborhood. This introduces a strong feed-back component in both types of systems. Yet, there are also differences. Usually, in cellular automata the updates occur strictly synchronously which is not true for a neural system. Many of the striking patterns produced by cellular automata under a synchronous update rule disappear when the rule is made asynchronous. This, however, does not mean that cellular automata with asynchronous update rules cannot produce interesting dynamics and structure. It only says that with respect to pattern formation, the same update rule may yield very different results in both cases (Novak & May 1992; Huberman & Glance 1993).

Another difference concerns the number of cells that affect the updating of a selected cell. In cellular automata, this is usually a small number of cells in the immediate vicinity of the selected cell. The update rule is usually identical for all cells, that is, it is translationally invariant. In contrast, a neuron is usually affected by several thousands of other neurons. In visual cortex, the majority of these are located in the vicinity of a cell, but there is also a significant fraction of long range connections. Also, the connectivity almost certainly is not translationally invariant. However, in some parts of the brain, including the primary visual cortex, it seems reasonable to expect an approximate translational invariance in some statistical sense, since the processing task looks spatially uniform. The required lengths of the connections between neurons may be strongly dictated by the nature of the information processing task they perform. Certain local filtering operations may well be feasible with short range connections plus a small fraction of longer range connections for coarsely coordinating the local filtering operations at distant sites (for a description of experimental results on lateral connections in the visual cortex, see, e.g., Gilbert & Wiesel 1989). Modeling the operation of such a system with cellular automata may suggest interesting variants of currently investigated automata. These could be obtained, for example, by introducing a small percentage of long range interactions in the update rule, or by using some "higher level stage" to change parameters in the local update rules such that the filter properties of the local filters are only correlated over finite distances.

4. DYNAMICAL SYSTEMS

Most of the systems considered in this book are characterized by a time evolution of some kind: a group of interacting agents evolves in time, a learning system proceeds towards its desired state, etc. The appropriate mathematical abstraction to capture this non-static nature is that of a dynamical system: a lawful motion of a state variable $x(t)$ in some state or phase space, governed by a set of differential equations

$$\dot{x}(t) = F(x(t)) \tag{8}$$

or by an update rule (a map)

$$x_{t+1} = G(x_t). \tag{9}$$

Even simple-looking dynamical systems may be capable of universal computation, in other words be able to emulate arbitrarily complex systems. However, one should not be too amazed on this. All that is needed is the demonstration that the dynamical system can emulate a few basic operations, e.g., conjunction and negation, and it must be sufficiently "modular" so that these operations can be combined in arbitrary ways. This suffices then to build the central processing unit of a computer, from which the rest follows. The time evolution of a dynamical system depends not only on a dynamics law like Eq. (8) or Eq. (9), but also on boundary conditions in space and in time. The notion of dynamical systems is so general that we cannot hope to cover all aspects relevant for prerational intelligence in this single section. Instead, we discuss some aspects within the other sections, and focus here on a few selected points.

A quite simple concept for dynamical systems is that of a fixed point $\hat{x}$, characterized by

$$F(\hat{x}) = 0 \quad \text{for differential equations,} \tag{10}$$

$$G(\hat{x}) = \hat{x} \quad \text{for maps.} \tag{11}$$

(In the following, we will use mainly the notation for differential equations; analogous equations will hold for maps.) Depending on whether nearby locations in state space will be attracted towards the fixed point, or driven away, one distinguishes attractors and repellors, or stable and unstable fixed points. The region in phase space consisting of all points which will ultimately be driven towards a particular fixed point is called the basin of attraction of this fixed point. Sometimes, the dynamics can be characterized by an ever decreasing function, with the attractors being minima of this so-called Lyapunov function. From the existence of such a Lyapunov function for a dynamical system one can deduce powerful statements about stability properties of the system (Cohen & Grossberg 1983). Unfortunately, there is no general recipe for constructing such Lyapunov functions; they have to be invented via some ingenious intuition about the system at hand.

Dynamical systems can exhibit several attractors simultaneously (multistability). Which of these will be chosen by the dynamics depends on the initial conditions of the system. The idea of multiple attractors, surrounded by basins of attraction, has been applied quite fruitfully to the concept of memory, which has implications for prerational intelligence. Hopfield pointed out that a simple neural network with symmetric Hebbian weights has an energy function, which decreases during asynchronous update steps (or at least remains unchanged) (Hopfield 1982, 1984). The energy function is chosen such that its minima are located at those points in state space which are to be memorized. Following this basic idea, numerous questions have been addressed using methods of statistical mechanics. How many patterns can a network store (Amit et al. 1985)? What amount of misinformation (noise) can be tolerated such that the pattern is retrieved anyway (in other words how large is the basin of attraction)? How fast are the patterns retrieved? By what modifications to the choice of connection weights or to the type of patterns can the storage capacity be increased? Mathematical theory, exploiting methods from the theory of spin-glasses, answered most of these questions and some additional ones; a review goes beyond the scope of this chapter (for a review, see, e.g., Amit 1989). We just conclude that in this example, that is, with regard to a particular type of model for an associative memory, parameters which are directly related to intelligence have been fully determined on a theoretical level.

Many systems do not have attractors, but instead follow a particular closed path through phase space; they remain on a limit cycle. Analogous to fixed points, limit cycles can be stable or unstable. Simple examples for relevant model systems with limit cycles are (nonlinear) oscillators like the van der Pol oscillator (as a model for membrane potential oscillations in a regularly spiking neuron (Fitzhugh 1961)) or the Wilson-Cowan-oscillator (as a model for periodic changes of excitation in two coupled excitatory and inhibitory neuronal populations (Wilson & Cowan 1972)). Further examples of biological systems which are related to prerational intelligence and which exhibit limit cycle dynamics are central pattern generators, which are discussed in more detail by Dean et al. (this volume). Again, a host of questions follows once a limit cycle is identified in a dynamical system. What is its periodicity? What is its basin of attraction? How do its shape and periodicity depend on parameters of the system?

When parameters of a dynamical system are changed in a continuous way, the resulting can vary in a qualitative fashion. Stable fixed points can become unstable and vice versa. A particularly interesting case is the transition from a stable fixed point to a limit cycle. Such phenomena, called bifurcations, are also amenable to mathematical investigation, and several generic bifurcations have been identified. Transitions between different attractors have been utilized, for example, to explain sudden changes in motor behaviour following a small change in external driving parameters (Schöner & Kelso 1988; Dijkstra et al. 1994; Schöner 2000).

In addition to fixed points and pure limit cycles, still more complicated dynamics can be found. One simple extension is a combination of two or more periodic motions with an irrational ratio of frequencies. In such quasiperiodic motion, the trajectory will never exactly return. Nevertheless, the dynamics appear to be quite regular to an observer (see Fig. 2). In contrast to all the types of dynamics we have described so far, and in contrast to the intuitive expectation that "simple systems" should exhibit "simple dynamics", particularly in the absence of noise, very complicated, irregular paths through state space have been observed in even the simplest nonlinear dynamical systems, like quadratic maps

$$x_{t+1} = ax_t(1 - x_t). \tag{12}$$

Even though trajectories generated by this or other nonlinear systems may even superficially resemble a stochastic motion, they are not stochastic, since they originate from a deterministic rule (the dynamics equation). Such tra-

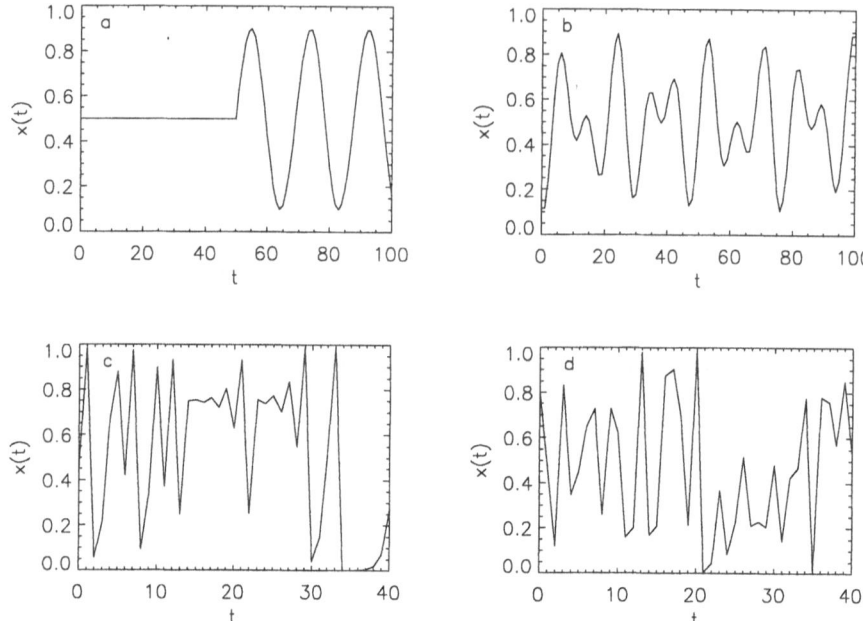

Figure 2: Illustration of qualitatively different types of dynamics. a) fixed point dynamics and oscillatory (limit cycle) dynamics, b) quasiperiodic dynamics (with the golden mean as the ratio of frequencies), c) chaotic dynamics (Logistic map $x_{t+1} = 4x_t(1-x_t)$), d) stochastic dynamics. The latter are not distinguishable by visual inspection, even though the one is purely stochastic (disregarding the fact that the series was generated by a pseudo-random number generator on a digital computer), whereas the other results from a low-dimensional deterministic process.

jectories have been called chaotic. A systematic characterization of chaotic systems and of the different routes into chaos is the major field of interest in nonlinear dynamics (for a review, see Schuster 1989; Strogatz 1994). In order to qualify as chaotic, several conditions have to be met. The power spectrum of the signal has to be continuous and not to consist of a discrete number of sharp peaks. The latter case would point to a time series resulting from a superposition of periodic signals. Even more important, the dynamical system has to be such that trajectories with neighboring initial conditions are driven apart in the course of time. In order to quantify this effect, one considers a starting point x_0 and a point x in its neighborhood,

$$x = x_0 + \epsilon \qquad (13)$$

One then relates the dynamics of x to that of x_0, by linearizing Eq. (8) about x_0,

$$\dot{x} = F(x) \approx F(x_0) + \frac{\partial F(x_0)}{\partial x}\epsilon = \dot{x}_0 + \frac{\partial F(x_0)}{\partial x}\epsilon. \qquad (14)$$

The eigenvalues of the resulting Jacobian $\partial F/\partial x$, averaged over the regions of phase space which are accessible to the dynamics, indicate at what rate deviations along the directions of the respective eigenvectors of the Jacobian shrink or grow. In case one of these so-called Lyapunov exponents is positive, small deviations between neighboring trajectories in the direction of the corresponding eigenvector will grow exponentially,

$$\langle x(t) - x_0(t)\rangle_{x_0(t=0)} = \epsilon e^{\lambda_i t}. \qquad (15)$$

As a consequence, two states of a chaotic system which differ initially by a minuscule amount will be completely different after a while. Decreasing the initial difference by a factor (i.e., multiplicatively) only adds to the time until the two states have diverged. Hence, the important parameter is the growth rate, and not the initial precision. This divergence phenomenon is called sensitive dependence on initial conditions. Even though such a dependence on initial conditions may seem opposed to the requirement of robustness, there are also advantages inherent to this growth pattern. Small differences can be better resolved in this way. Considering also the possibility that the type of dynamics (chaotic/nonchaotic) can be controlled by a small number of parameters, possibly a single one, the notion of an "adaptive magnifying glass" comes to mind. This has been exploited, for example, to achieve the control of chaotic systems with tiny control signals (Ott et al. 1990). A quick separation of nearby states due to chaotic dynamics has also been suggested as a means of quickly and robustly desynchronizing oscillatory elements that ought to be kept separate (Hansel & Sompolinsky 1992). The fast generation of ever new neuronal firing patterns in a neural system with chaotic dynamics has also been argued to be responsible for the fast course of olfaction, in particular, and of perception, in general (Skarda & Freeman 1987; Freeman 1991).

Let us now turn to the problem of how much of phase space is occupied by a dynamic system. A fixed point occupies only a point in phase space ($d = 0$), a limit cycle occupies a line (d=1), whereas the dimension of the phase space itself is determined by the number of variables describing the system. For chaotic systems, the trajectories are restricted to regions of phase space, the dimensions of which are noninteger. This gives rise to the notion of a "strange attractor". Several types of dimensions can be defined, for example, the Hausdorff-dimension, the information dimension or the correlation dimension. The derivation of relations between these dimension measures, the Lyapunov-exponents and further quantities, which help characterize the system, has been a major research topic in nonlinear dynamics.

All these concepts (and a few more which will be introduced soon) can be fruitfully utilized for an analysis and description of prerational intelligent systems, when one approaches a chaotic system from the output signal. If signatures of chaos can be found and quantified in an irregular signal, then one knows it has been generated by a rather simple, albeit nonlinear, dynamical system. It is not just random. Furthermore one can proceed to identify quantitative properties of the underlying dynamics. Irregular signals, which can be subjected to such an analysis, are abundant in systems to which we attribute prerational intelligence: from spike trains up to EEG-data (for an overview of such activities, see, e.g., Degn et al. 1987; Basal 1990, or Glass 1994). Behavioural data from a robot have also been considered in this context (Smithers 2000).

A central aspect of the analysis of experimental data stemming from a possibly low-dimensional dynamical system is that one cannot observe all relevant variables of this system, particularly if the relevant variables are not known. Instead, time series of experimentally accessible variables – typically just one variable – are measured. One variable by itself does not suffice to span a phase space large enough to embed the attractor we would like to characterize. However, using the method of delay coordinates, a one-dimensional time series $r(t)$ can be transformed into a new time series $s(t)$ in an m-dimensional phase space

$$\mathbf{s}(t) = (r(t), r(t - \tau), r(t - 2\tau), ..., r(t - (m - 1)\tau)). \qquad (16)$$

In this reconstructed phase space, essential properties of the underlying dynamical system, which gave rise to the observed time series $r(t)$, are preserved (Takens 1980). This applies in particular to the attractor dimension which can now be measured in the reconstructed space (Grassberger & Procaccia 1983). A more ambitious goal than to estimate quantitative properties of the data would be to identify the form of the underlying dynamics from the data. Unfortunately, the art of deciphering time series with regard to equations has not yet led to many fruitful results.

For the case of the already mentioned EEG-signals, attractor dimensions have been determined during various stages of sleep. It has been found that the deep sleep phases correspond to a system with markedly lower dimension than the REM-phases. Similar decreases have been observed in EEG-data during epileptic episodes of patients (Babloyantz & Destexhe 1986).

Both results suggest that a "productive" state of mind is associated with rather high-dimensional dynamics, whereas low-dimensional states occur in idle or even destructive periods. With regard to the identification of a high-dimensional ($d > \approx 3 - 4$) nonlinear deterministic system, it should also be mentioned that substantial amounts of data are necessary to reliably distinguish it from a stochastic system.

If a neuronal system is considered as a dynamical system, the concepts of dynamical systems theory can be applied to the system as a whole, or to the individual elements. The first approach was the basis for the notion of memory contents as the attractors of a neural network. It has also been used in phenomenological theories put forward by Gregor Schöner and colleagues, where macroscopic ("phenomenological") variables enter into the dynamics equations. If the dynamics are primarily ascribed to the individual constituents which, however, are coupled, then the additional question concerning properties of the overall dynamics has to be raised.

In the most general case of a fully connected, asymmetric neural network a huge range of dynamical phenomena can occur even if the individual neuronal elements are simple. Therefore, it is hard to make any general statements. However, asymmetric networks with suitably specialized weight patterns may allow a more specific characterization of their dynamic properties. One tractable case concerns networks with random connectivity. Another class of asymmetric networks results when a given fraction of weights in a symmetric network is cut. Many properties of such networks are obtained correctly from the corresponding symmetric network by adding a noise term. There are many further types of asymmetric networks that may be considered. An important theoretical question then is how to choose the connectivity to implement a desired dynamical behavior. Chaos of various forms has also been observed in coupled map lattices, systems where the individual elements themselves are capable of chaotic dynamics. In order to fruitfully employ such systems for information processing tasks, not only the coupling schemes, but also the boundary conditions will have to be chosen in a suitable way.

As was already mentioned repeatedly in this chapter and the chapter on biological issues, there is a severely restricted form of collective dynamics that is well suited to information processing, in particular with regard to the flexible and simultaneous binding of several entities. This is synchronicity within a subpopulation of elements in a neural system. From the mathematical point of view, in particular from the point of view of dynamical systems theory, this coding scheme raises many questions, which have to be addressed in modeling studies of neural networks with synchronized dynamics. When does synchronicity arise (i.e., under what stimulus conditions)?

How fast does it arise (fast enough to be usable for information processing)? To what degree does it depend on properties of the individual constituents of the system (can frequency deviations of local oscillators be tolerated)? How does the synchronization depend on the strength and delay of the couplings? An analytic investigation of these questions is particularly difficult if the constituents are not exchanging signals all the time (phase coupling), but instead are only exchanging pulses at selected points in time (pulse coupling). The latter scenario is easier to identify with individually spiking neurons and collectively bursting groups of neurons. Numerically and analytically, some results on these problems have already been achieved (a large number of investigations have been published recently: see, for example, Kuramoto 1991). Other problems are still to be resolved. One of the latter is the question of how periodic signals can be synchronized if they are coupled with a delay which may exceed a quarter of the oscillation period. Coupling two oscillators in such a way should yield an antiphase-relation. Yet, the inter-hemispheric periodic synchronization observed in cat visual cortex (Engel et al. 1991) involves long distance projections, the delays of which amount to about a quarter period ($T/4$=6.25ms at 40Hz) or more.

5. OPTIMIZATION

Another immediate association we have when thinking about ways to formalize aspects of prerational intelligence is that of optimization of some resource or cost. Optimization principles have proven to be flexible and powerful foundations from which laws and properties of many systems in nature can be derived in a concise and parsimonious way. A prime example is the derivation of classical mechanics from the principle of least action. In chemistry, chemical reactions proceed to minimize the Gibbs potential. In biology, evolutionary progress is usually viewed as increasing some fitness function of a species. Imitation of this evolutionary process in the engineering domain has led to an interesting optimization strategy, which has successfully been applied to a number of technical problems (Rechenberg 1973; Holland 1975; Goldberg 1989) (see below). Considering this broad range of applicability of optimization principles, we can be quite confident that many aspects of prerational intelligence might be derivable in a similar way, that is, as consequences of minimizing suitably chosen cost functions.

In the context of behavior of agents the notion of cost functions which are to be optimized is related to the notion of goals. Depending on the representation of goals in the system, one can distinguish goal-achieving systems (which identify a goal once it is achieved, but which are not guided towards the goal), goal-seeking systems (which seek a goal, even though it is not represented inside the system), and goal-directed systems (see *Philosophi-cal Perspectives on Prerational Intelligence*). The concept of optimization presumes that the cost function is repeatedly evaluated during the course of minimization, or even that the explicit knowledge of the cost function is ex-ploited, for example, when computing gradients. We can therefore regard systems which optimize cost functions as goal-seeking or goal-directed, but not as goal-achieving.

In the following we would like to focus on a few particular examples of optimization processes and cost functions which are related to the topic of this book. The first example is concerned with "optimal" decisions. Clearly, an intelligent system, which behaves rationally, tries to make as good de-cisions as possible. The term "rational" is used here in the sense that the decision is based on an evaluation of the cost function only, as opposed to irrational decision making which involves additional factors. This distinc-tion is different from the one drawn between prerational and rational intel-ligence in the general context of this book. A simple mathematical frame-work to connect the costs of individual (mis-)classifications and decisions is provided by Bayesian decision theory (for an introduction to Bayes theory, see, e.g., Duda & Hart 1973). This issue has been pursued further in the field of game theory, where more complicated decision situations are inves-tigated (Selten 1982). As an aside remark we can note that experimental studies seem to contradict the assumption of purely rational decision mak-ing by humans (Davis & Holt 1992). Bayesian decision theory is also an accepted procedure in the fields of signal detection and pattern recognition, tasks which can be regarded as decisions on the prerational level.

To apply the Bayesian approach, situations are described by some "fea-ture vector" $\mathbf{x}$, decision options are indexed by j. Each decision is based only on the current situation $\mathbf{x}$, and is viewed as independent of previous situations. In Bayesian decision theory, the task of making a decision is formally described as a classification task: given a range of possible obser-vations $\mathbf{x}$, stemming from a discrete number M of classes j, the objective is to assign to each $\mathbf{x}$ a class label $j = c(\mathbf{x})$ such that some measure of the cost of making misclassifications is minimized. In order to be able to specify such a measure, the existence of two sets of probabilities is assumed. The first set $p(j)$, $j = 1, ...M$ of probabilities describes our a-priori assump-tions about the probability of each class occurring. Furthermore for each class j the conditional probabilities $p(\mathbf{x}|j)$ give the distribution of values of $\mathbf{x}$, provided class j is present. Through these quantities a-priori knowledge about the environment can be incorporated into the decision process. Finally, the matrix C_{ij} describes the cost of misclassifying $\mathbf{x}$ from class j into the (wrong) class i. The proper assignment of values to C may be highly context

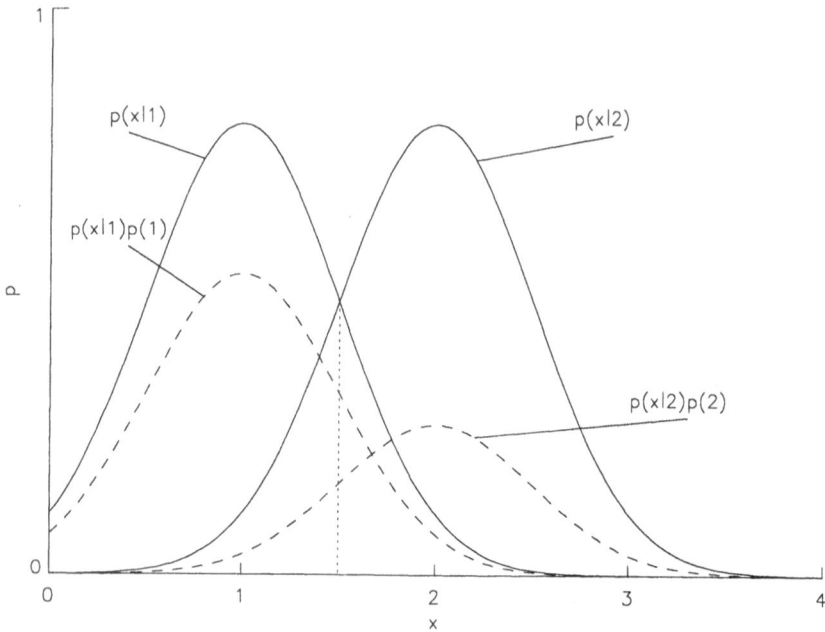

Figure 3: Illustration of decision boundaries for the classification of data points x into two classes 1 and 2. If class j is present, the data points x are distributed according to $p(x \mid j)$. Here, for both classes the distribution is assumed to be of Gaussian shape with identical width. If both classes are equiprobable, and the costs of misclassification are identical, the optimal Bayes decision boundary is located at the mean value of the two distribution centers (dotted line). If the overall probabilities $p(j)$ for the classes to occur differ (in this example, $p(1) = 2/3$, $p(2) = 1/3$), or if the shape of the distributions $p(x \mid j)$ is not Gaussian, the Bayes rule results in a decision boundary which differs from the mean of the class mean values; a data point x is put into that class j the product $p(x \mid j)p(j)$ of which is largest. In a graphical representation, the decision boundary is located at the intersection of the corresponding curves (i.e. at about $x = 1.7$ in our example).

dependent (compare, for instance, the costs of misclassifying a predator for a prey or vice versa).

All costs of misclassifications are positive, and the costs of correct classification vanish (i.e., $C_{ii} = 0$). Given these quantities, we can now ask, what choice of the classifier $c(\mathbf{x})$ minimizes the following average cost of misclassification,

$$E(c) = \sum_j p(j) \sum_i C_{ij} \int_{x \in X_j \wedge c(x) = i} p(x|j) \, dx. \qquad (17)$$

Minimizing $E(c)$ with respect to $c(\cdot)$ is an optimization task. Its solution leads to Bayes' decision rule. In many cases, $C_{ij} = C = const$ for all $i \neq j$. Then, a simple solution arises, picking that class j for which the product $p(\mathbf{x}|j)p(j)$ is maximal (see Fig. 3).

The role of the Bayesian approach is not limited to optimizing discrete behavioral decisions. It also provides a theoretical framework for how to extract information from sensory signals in order to infer properties of the environment. Some of the consequences of this approach have been extensively considered in vision. Here, the geometrical properties of objects have to be derived from the intensity distributions on the retina (or a CCD-chip in artificial systems). The notorious difficulty of this problem makes it one of the prime examples for a process which requires intelligence to be solved, but the solution of which cannot yet be described in a set of rules, neither by insight (into biological systems) nor by construction (of artificial systems). The difficulty is related to the problem that most of the geometric properties of objects are grossly underspecified by the observable 2d-intensity distribution (at least for static images). Therefore, additional constraints, like smoothness and edge-discontinuities, are imposed – a method which goes under the name of "regularization" of an otherwise ill-posed problem (Poggio et al. 1985). If the constraints are imposed via a suitable choice of the probabilities p_j and $p(\mathbf{x}|j)$, respectively, the approach is called using random Markov fields (German & German 1983).

The Bayesian framework is not the only method where optimization is relevant for the study of prerational intelligence. Earlier, we have pointed out that many forms of prerational intelligence seem to be emergent properties in systems of many nonlinearly interacting constituents or agents. In the majority of cases, even strong simplifications of the properties of these constituents and their interactions lead to model systems that are too complex for analytically deriving global properties. Instead one has to resort to simulations. This does not come as a surprise, because it is the very nature of emergent (global) properties to preclude any simple derivation from the properties of the constituents. However, it has the consequence that systems which are to exhibit emergent properties which meet some given (e.g., by a designer) specifications can only be constructed by some iterative optimization approach. Therefore, to use the creation of suitable models as a means to identify mechanisms which allow certain properties to emerge, or to exploit such mechanisms for engineering purposes, suitable optimization methods have to be developed and applied.

In principle, many optimization methods can contribute to this goal. In the following, however, we will focus on a subset of methods that are characterized by interesting analogies to naturally occurring processes, namely thermal annealing, learning and evolution.

The first method, *simulated annealing*, derives its name from its similarity to the physical process of annealing during the cooling of a metal. During the annealing phase the atoms in a metal adjust their positions such that the free energy of the metal piece becomes minimized. In the simulated version of annealing an optimization task is formulated in terms of a system of interacting, simple constituents, each characterized by a small number of states, such that minimization of some "energy" E resulting from the interaction of the constituents defines the optimization goal. The optimization process itself is then defined by some suitable (usually discrete) dynamics for the states of the constituents. At each time step, the change ΔE of the total energy is considered, when a randomly selected constituent changes its current state from s to a randomly selected new state s'. If this would lower the energy (i.e., if $\Delta E < 0$), the considered state change is accepted, that is, carried out. Otherwise, it is accepted only with some probability p that decreases from $p = 1$ for $\Delta E = 0$ to $p \to 0$ as $\Delta E \to \infty$. A sequence of such steps constitutes a search in the state space of the entire system such that changes that lower the energy are preferred, but occasional increases of E are tolerated, thereby allowing escape from local minima (Metropolis et al. 1953). The range of positive values ΔE that are accepted with some appreciable probability is controlled by a "temperature parameter" T. For very large values of T, any change is about equally probable. The search is totally random, exploring the entire state space. As T approaches zero, the only remaining changes which are accepted are those for which E decreases. The search then becomes restricted to the basin of attraction of the nearest local energy minimum. Usually, T is chosen large initially and then gradually lowered, thereby interpolating between the two extremes just described (Kirkpatrick et al. 1983). Under such a "cooling schedule" many optimization problems exhibit phenomena that are very similar to the phase transitions observed when cooling a gas or a liquid. As in these physical counterparts, slow cooling in the vicinity of phase transition regions is essential for obtaining good optimization results.

Simulated annealing can be applied to the construction of neural networks to study how tasks like data compression, association and non-linear mappings can be solved by such systems. One major approach has been termed the "Boltzmann machine" (Hinton & Sejnowski 1986). Here the optimization parameters are the weights of a symmetric network, and the optimization goal is to minimize the difference between corresponding pair correlations when the output units are clamped to the desired output pattern and when they are free running. However, due to the large state space, the construction even of medium-sized networks by this approach requires large computational resources. This has motivated a search for optimization schemes that can operate more efficiently for this task. One recently adopted strategy has been to restrict the state space of the search by imposing constraints on the allowed activity patterns of the units (Kappen 2000). Another approach combines simulated annealing with gradient-based methods, such as the backpropagation algorithm (Rumelhart, Hinton, & Williams 1986).

Using gradient information can speed up considerably the convergence to a minimum. However, gradient information can be obtained easily only for pure feed-forward networks. The (error-)backpropagation algorithm (Rumelhart et al. 1986) and its many variants are well-known schemes for the necessary computations. For recurrent networks, the computation of gradient information becomes much more involved, and the convergence properties of the entire optimization process become even more obscure. Despite these problems, gradient-based approaches to the construction of neural networks with specified properties have offered a new approach to the study of how such systems may function. In an artificially created network all cell properties can be examined at will, and their dependence on system parameters can be investigated. This has contributed insights to many issues, in particular with regard to how information may be represented in the activity and the connectivity of neural systems. Good summarizing examples can be found, e.g., in Churchland and Sejnowski (1993).

Optimization approaches have also been applied to tune the behavior of one "agent" or a collection of interacting "agents" in some simple environment. For this, genetic algorithms have proven valuable. It has been shown that rather sophisticated control behavior, such as balancing a double inverted pendulum, can be achieved in this way.

The representation of computations in the form of networks of simple computing elements with adaptive connections has motivated the investigation of many further, hitherto unexplored forms of optimization. The goals of these optimization schemes are manifold: speeding up convergence, achieving better generalization, improving noise tolerance and saving computational resources, to name a few. Among the more prominent approaches that have been explored are methods of "weight pruning" by adding suitable terms to the cost function, methods to incrementally build networks, or particular ways of prestructuring that have interesting relationships to earlier statistical approaches and that lend themselves to the application of optimization methods, such as the EM-algorithm.

6. LEARNING

Learning has been defined as "any relatively permanent change in behavior produced by past experience" (*Encyclopedia Brittanica*) and is an important aspect of intelligence, rational or prerational. In the last 100 years, psychological research has revealed many different types of learning. Some of them, such as cognitive learning, clearly must be attributed to the rational level of intelligence. Others, in particular classical conditioning and operant conditioning as observed in many species, appear to affect processing at the prerational level. Therefore, a clarification and understanding of the notion of prerational intelligence has to take the issue of learning into account (for a discussion of learning from different points of view than the mathematical, see Dean et al., this volume). Mathematical theory can contribute here by developing models and techniques which describe and evaluate learning processes and their success. Such models offer frameworks in which experimental results can be interpreted. New experiments may be suggested and their outcome could be predicted. In addition, models allow analysis of learning from a computational viewpoint and from the viewpoint of information processing, with the goal of achieving a unified view of different learning processes and their properties. Finally, mathematical theory can provide valuable guidance for the design of artificial systems with learning capabilities.

There are various ways that learning might be formalized mathematically. Of these, we focus in the following on two major ones: *learning as optimization* and *learning as processing of information*.

Learning as optimization considers learning as having the goal of adjusting certain "adaptive parameters" such that some cost function becomes minimized. Following a standard notation in the artificial neural network literature, we denote the set of these parameters by **w**. While the cost function characterizes – explicitly or implicitly – *what* the learner tries to achieve, it does not characterize *how* the learner is to proceed towards his goal. Therefore, one also needs to specify a process – the "learning rule" – that asserts how an appropriate parameter setting **w** is achieved. Within this approach, learning is usually modeled as an iterative process, consisting of a sequence of "learning steps", with each learning step making small changes $\Delta\mathbf{w}$ to the adaptive parameters to improve the minimization of the cost function. This rough sketch of a learning process suffices to raise several questions which are typical for the mathematical perspective on learning. When and how rapidly does a learning process converge? What are the limits on the asymptotic performance of the learner? What approximations will result from limiting the learning to a finite number of steps?

An important characteristic which distinguishes different possible learning rules is the type and amount of information they require in order to perform a learning step. One piece of information a learner can get is some input signal **x**, coding a state of the environment. Depending on the learning situation, **x** is often called a feature vector. The learner is to do something on **x** and to come up with a response $c(\mathbf{x}, \mathbf{w})$, depending on its parameter set **w**. The quality of the response can be evaluated by a cost function E.

Learning rules can now be subdivided according to the following scheme: supervised learning rules require knowledge of the (desired) solution y in response to x, which would minimize the cost function. At the other extreme, unsupervised learning is based solely on the inputs x, without any explicit information about its relation to a cost function. A third form is reinforcement learning which provides the learner after some, but not necessarily all learning steps, with information about the value of the cost function (but not about desired solutions).

In supervised learning, the cost function frequently is of the form

$$E_i(w) = (y_i - c(x_i, w))^2, \tag{18}$$
$$E(w) = \langle E_i(w) \rangle. \tag{19}$$

Note the similarity of this expression with the cost function of Bayesian decision theory, when we regard the squared distance in Eq. (18) as the cost of misclassifying y_i for $c(x_i, w)$. In fact, if the task of the learner is a classification task, finding (an approximation to) the optimal Bayes classifier $c(\cdot)$ becomes equivalent to finding the (asymptotic) solution to a supervised learning task. However, the tasks that are solvable by supervised learning are more general.

An important property of a learning rule is how it determines which changes are effective for the minimization of the cost function, and which changes are irrelevant. This is known as the problem of *credit assignment* and its solution is central to devising a good learning algorithm. For supervised learning rules, with the dependence of the cost function on the parameters w explicitly known, the gradient ∇E of the cost function with regard to w can often be computed. Then the credits of the individual parameters are assigned proportional to this gradient, and a (*gradient descent*) learning step is performed according to

$$\Delta w = -\epsilon \nabla E(w). \tag{20}$$

For many cost functions that are expectation values over all past and future trials, such as, for example, (19), $E(w)$ is not known exactly. This gives rise to the introduction of approximations that estimate E on the basis of the current or a number of recent trials. One frequent idealization is to assume statistical independence between trials. In the case of (19), Eq. (20) is replaced by

$$\Delta w = -\epsilon \nabla E_i(w), \tag{21}$$

where $E_i(w)$ can be evaluated after each individual trial. For sufficiently small learning rate ϵ, one has

$$\langle \Delta w \rangle = -\epsilon \nabla E, \tag{22}$$

that is, the mean value $\langle w \rangle$ follows the original learning equation (20).

One example of these supervised learning schemes, error backpropagation in multilayer-perceptrons (Rumelhart et al. 1986), has gained enormous popularity in the field of artificial neural networks. Here, we just comment on some aspects of backpropagation, which are related to our topic of prerational intelligence, and refer the reader for a more thorough treatment to the previous chapter in this volume. First one has to note that the type of learner employed in this scheme, a multilayer-perceptron, is a neural network, that is, it resembles to some degree the hardware of systems exhibiting prerational intelligence. However, two words of caution have to be said. First, the analogy between multilayer-perceptrons and biological neural networks has several weaknesses. The degree of precision required in synaptic transmission is unrealistically high, particularly during the learning phase (16 bit resolution for the weights during learning, 8 bit during subsequent operation (Debenham & Garth 1988). With regard to the weight adaptation, one has to note that it is not of Hebbian or related form. Even though on the presynaptic side the amount of change is proportional to the excitation level, on the postsynaptic side it is not. Instead the size of some error signal is decisive, which has to be retrogradely transmitted from some omniscient teacher, perhaps even across several synapses. No such physiological mechanisms have been identified (yet). Furthermore, one of the general criticisms of artificial neural networks applies, which dwells on the fact that in artificial networks the weights can freely change signs during learning, and that an individual neuron can have excitatory as well as inhibitory connections to other neurons. Such flexibility has not been observed in biological neural networks, where each neuron can only either excite or inhibit other neurons, but not both.

Our second point directly relates to the degree of (prerational) intelligence attributed to backpropagation. The degree of intelligence attributed to a learning system depends to some amount on the speed of progress during the adaptation phase. Backpropagation, however, is notorious for long convergence times. This weakness is inherent to the simple gradient descent which is employed in this learning rule. The parameter vector **w** can remain for long times on flat plateaus in the energy landscape (with very small gradients), it can even become trapped in local minima, or it can oscillate at the bottom of elongated valleys in the energy landscape. Many theoretical and empirical contributions have been concerned with possible speed-ups to overcome these problems (see, e.g., Fahlman 1988; van der Smagt 1994). One exemplary modification of the original learning rule is the choice of successive learning steps in the direction of mutually conjugate gradients in order to avoid oscillations in elongated valleys. The problems with local minima and plateaus can be remedied by introducing noise in each adaptation step, for example, by random sampling instead of systematic sampling. This creates a chance for occasional "uphill"-motions in the cost function, and in this way provides an escape mechanism in cases where the learning procedure gets stuck.

At the other extreme case of learning rules, in unsupervised learning, the learner has neither information about a desired solution associated with a particular input x, nor about the value of the cost function, which is associated with its response. A desired solution may not even exist. All the learner can (and should) achieve is to detect certain regularities in the environment. Obviously, which types of regularities are extracted has to be implicitly contained in the learning rule.

Even though a cost function is not explicitly utilized during learning, it would facilitate the analysis of the resulting learner if a cost function were known which is minimized during the learning process. Existence of such cost functions would also help to compare different learning rules, to evaluate the final success of the learning, and to evaluate the rationality of process inherent to the learning scheme. It is an interesting theoretical question to ask in which cases the considered learning process optimizes some cost function, even though it does not utilize it during learning. By virtue of an analogy between orientation preference maps in visual cortex, with left-handed and right-handed pinwheels, and electrostatics, with positive and negative charges, a cost function has been identified for this kind of visual map, which is analogous to the electrostatic potential (Wolf et al. 1994). This cost function takes on rather low values in experimentally measured maps, and it is found to decrease during adaptive self-organisation processes modeling the ontogenesis of such maps. Interestingly, the evaluation of this cost function, which is completely unrelated to the learning process, indicates that learning still makes progress even in the final stages of simulated self-organization processes when other means of evaluation, like a visual comparison between model map and experimental map, do not show any apparent difference. A subsequent question is whether this learning process can be formulated as a gradient dynamics using some cost function (or potential in the terms of physics). It turns out that there are cases where the answer to this question is negative. One example is the self-organization of topology-preserving maps according to Kohonen's rule. This process has been shown not to be derivable as a gradient dynamics from any cost function (Erwin et al. 1992). Other learning processes for the same task, such as the elastic net algorithm, do follow the gradient of some cost function (Durbin & Willshaw 1987).

Unsupervised learning schemes like Kohonen's self-organizing feature map have been widely investigated with regard to self-organization phenomena in nervous systems (see also the discussion of pattern recognition in Section 3). Many feature mappings in various sensory modalities appear to be learned in such a way (Ritter & Schulten 1986; Martinetz et al. 1988; Obermayer 1990; Wolf et al. 1994). Moreover, the reorganization of maps following a change in the statistics of the peripheral inputs (e.g., as a consequence of deafferentiation) is among the best-investigated examples of ongoing plasticity in nervous systems (see, e.g., Kaas et al. 1983). Many of the effects observed in such experiments can also be explained on the basis of unsupervised (re)learning schemes (Ritter & Schulten 1986). As a consequence of this adaptability and flexibility, such systems can certainly be considered to show prerational intelligence. In addition, the topography inherent to some unsupervised learning schemes provides robustness when isolated elements become dysfunctional. In this case, a neighboring element will be activated instead of the corrupted one, and due to the topography, the input information which is encoded by this surrogate will not be too misleading.

Between the two extremes of supervised and unsupervised learning is an intermediate type, reinforcement learning. In reinforcement learning, a response of the learner to an input signal x_i produces only a reward value E_i, without any information about a correct or target response y_i. As in unsupervised learning, gradient information ∇E_i is no longer available. In the general case, instead of a gradient directed search some random search procedure has to be employed, guided solely by the observed values E_i. A particularly important learning situation of this kind requires a sequence of actions to be found that only in proper conjunction lead to reinforcement at the end of the sequence (consider a mouse in a maze, which will find food only after a particular sequence of left/right turns). For this type of learning task, reinforcement learning algorithms have been developed that can be viewed as stochastic approximations of dynamic programming. The idea behind these approaches is to consider each trial as the selection of some action c, given the current state x. The new nomenclature "state" for the input to the learner results from the learning situations in which such schemes are usually employed; the state x plays the same role as the input vectors x of the previously discussed learning schemes. The selection of an action is based on an estimation procedure for the following cost function:

$$E(x) = r(x, c) + \gamma r(x', c') + \gamma^2 r(x'', c'') + \dots \tag{23}$$

Here, x denotes the present state of the system, $x', x'' \dots x^{(n)}$ denote the successor state $1, 2 \dots n$ steps into the future. $c, c' \dots c^{(n)}$ denote the associated actions, and $r(x, c)$ is the cost of choosing action c in state x. $r(x, c)$ includes the change in the cost function due to a possibly decreased distance to reward states, as well as costs due to actually performing the action (consider, e.g., energy used). The prefactor $\gamma < 1$ has the effect of "discounting" costs that are more distant in the future.

As in the case of supervised learning, $E(x)$ is not known explicitly during the learning process. Again, one resorts to a stochastic approximation by introducing a function $Q(x, c)$ that is an estimate of $E(x)$, provided the next action has already been decided to be c (Watkins 1989). If action c has been executed and led to a cost of $r(x, c)$, this information can be used to update the current estimate for $Q(x, c)$ according to

$$\Delta Q(x, c) = \epsilon(r(x, c) + \gamma E(x') - Q(x, c)). \tag{24}$$

The idea behind this equation is that now the first term in Eq. (23) is known to be $r(x, c)$, the remaining terms in the sequence being given by $\gamma E(x')$. The iteratively refined estimates $Q(x, c)$ are used to select an action at each step. As a general guideline, one selects that action c that leads to a maximal value of $Q(x, c)$. However, as in the case of supervised learning, introducing some noise leads to better exploration of the space of solutions. Here it is not only advantageous, but essential for the success of the method.

The just described method of Q-learning resembles some aspects of the operant and classical conditioning paradigms in psychology. However, whereas in classical conditioning a correlation between the stimulus and the response is learnt, and in operant conditioning a correlation between the response and the reinforcement function, the Q-matrix in Q-learning connects all three. $Q(x, c)$ approximates the reinforcement value $E(x)$ for a particular stimulus x, provided the response was c.

Considering that the learner bases his responses on an estimation of the reinforcement function which at least partly is explicitly given only at the end of a whole sequence of steps, this learning scheme shows anticipation. This is an indicator of intelligence as is the directed variability inherent to the approach (compared to purely stochastic variability).

At this point we should also comment on characteristic features of learning curves and their interpretation. The absence of plateaus and sudden drops in the learning curves of cats learning to escape from a box by pulling

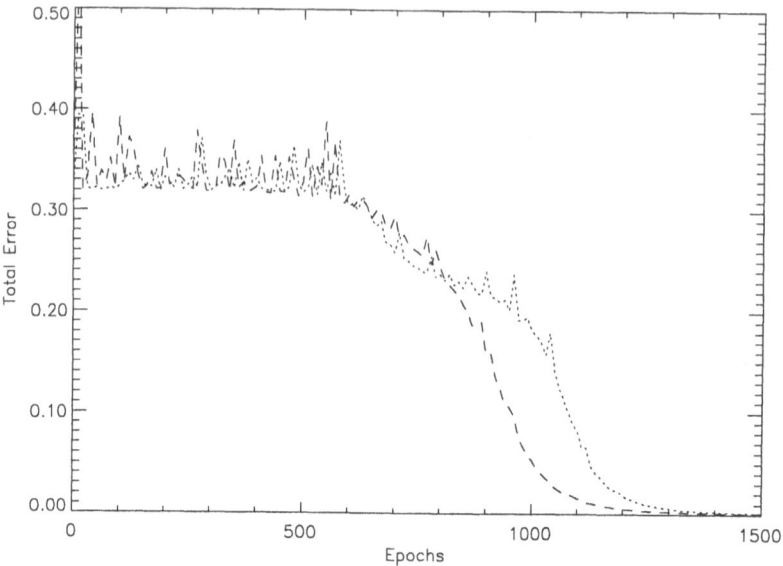

Figure 4: Learning curves for XOR-Problem in a 2-2-1-multilayer perceptron (MLP (dotted curve) and 2-10-1 MLP (dashed curve). Both curves initially remain on a plateau, before they suddenly drop to small values of the total error E. The patterns were presented in random order during learning, and the total error does not necessarily always decrease during learning for this reason.

a lever led Thorndyke to assume that the cats did not gain insight into the role of the lever, but instead learned a stimulus-response-bond association between a stimulus and a motor action (Thorndyke 1966). Inverting this argument, one could argue that plateaus and sudden drops in learning curves correspond to insights gained during a learning process, and should, consequently, be taken as signs of intelligence involved. Learning curves can easily be obtained from all learning rules which involve cost functions, whether they are utilized in learning or not. Investigating such curves, plateaus and drops can be found, for example, when learning the famous XOR-problem with multilayer-perceptrons and backpropagation learning (see Fig. 4).

Following the psychologist's arguments about the drops in learning curves, one would conclude from Fig. 4, that in both cases the network deciphered some sort of rule about the XOR. Yet, as is shown in Fig. 5, the investigation of the hidden layer activities does not always allow an easy interpretation of what the network does in terms of simple rules (in fact, it rarely allows such an interpretation; as an example of a more ambitious interpretation of the hidden layer activities and connectivities, see Sejnowksy & Rosenberg 1987). The smaller network apparently gained "insight" into the problem at hand, since its nodes fire according to a rule, which seems to grasp the essence of the computational problem. The larger system apparently did not

Pattern (1,1)

Pattern (1,0)

Pattern (0,1)

Pattern (0,0)

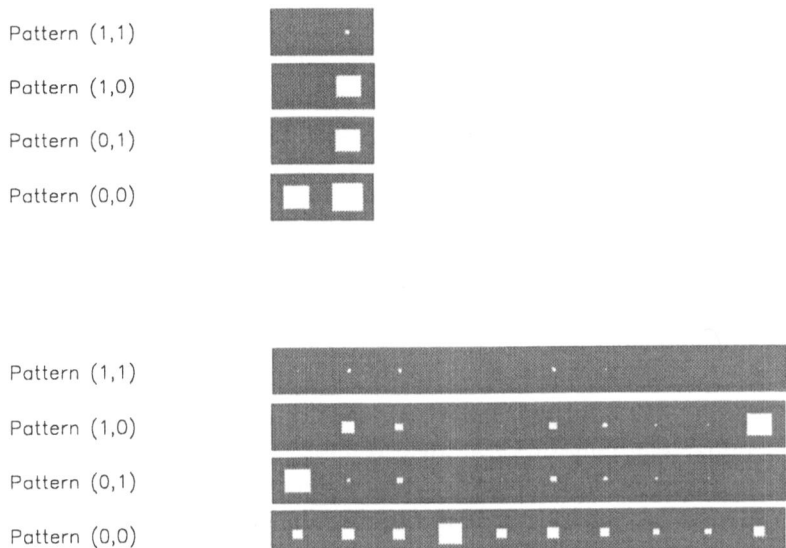

Pattern (1,1)

Pattern (1,0)

Pattern (0,1)

Pattern (0,0)

Figure 5: Hidden layer activities for a 2-2-1 (upper part) and a 2-10-1 (lower part) multilayer perceptron trained on the XOR-problem (see also Fig. 4). Responses are shown for each of the four patterns constituting the XOR-problem, respectively. The level of hidden node activity o, $0 < o < 1$ is indicated by the size of the white squares: vanishing size: $o = 0$, maximum size: $o = 1$. The 2-2-1 network seems to have detected a "rule": the left hidden units fires only when neither input channel has a signal, the right hidden unit is inhibited when both channels have input signals. A firing of the right unit, and non-firing of the left indicates the presence of a pattern with a positive XOR-relation. In contrast, in the 2-10-1 network, three units seem to have specialized for particular patterns, and if either the rightmost or the leftmost unit fires, a positive XOR-relation between the input channels is indicated.

gain "insight", but devoted some of its hidden units to learning particular patterns by heart. The question arises as to whether one should assign the smaller system a higher degree of intelligence than the larger on the basis of this difference. At least we can summarize that a systematic analysis of learning curves, and a comparison to interpretation categories used in psychology, is a worthwhile enterprise, which might also give some insights into the learning of rules.

The optimization approach is mathematically a very convenient starting point for describing the aim of learning and for casting such aims into learning rules. However, as already indicated at the beginning of this section, there is a different viewpoint that emphasizes the *information processing aspect* of learning. This may make it easier to relate to psychological studies of learning. It is also more general, since it does not necessarily restrict learning to optimization.

What does it mean to view learning as information processing? Information comes into play on at least two different levels. The first level concerns the choices made by the learner. Intuitively, learning is accompanied by a gradual removal of uncertainty about "the right" action to choose. One way to mathematically grasp this uncertainty is to assign to each option x a certain probability $p(x)$ that measures how likely the learner will opt for this action. Information theory then considers the quantity

$$S(p) = -\sum_x p(x) \ln p(x) \tag{25}$$

for measuring the degree of uncertainty reflected in these probabilities. S is also known as "information entropy" and forms the basis for information theory. Learning then should be accompanied by a gradual decrease of this entropy. Therefore, S can be viewed as a particular cost function from the optimization point of view. Indeed, replacement of the quadratic cost function (18) in multilayer perceptrons with the entropy (25) has been investigated and found to lead to superior results in some instances (Solla et al. 1988; Bergman et al. 1991).

There is a second level where information has to be considered. The learning process itself can be viewed as gradually removing uncertainty about the "right" parameters w of the learner. Again, this can be described formally by a set of probabilities $P(w)$ that measure how likely the correct parameter values will be w. As before, we can consider an analogous information entropy

$$S(P) = -\sum_w P(w) \ln P(w) \tag{26}$$

for the uncertainty occurring on that level.

Both information measures or "entropies", $S(p)$ and $S(P)$, are related. Initially, $S(P)$ reflects the diversity of the learners structure, $S(p)$ reflects the diversity of the learners actions. The form of the exact relation between both types of diversity depends on the circumstances, and learning should gradually reduce both.

This formulation leads to several new questions: what kind of information is needed to learn? How can a learner distinguish relevant from irrelevant information? How do factors such as the amount of available information and its encoding affect the efficiency of learning with respect to time or computational resources? How is this related to the difficulty of a learning task? How can the goal of the learning process be characterized in information processing terms? How do issues such as generalization and discrimination come in?

With regard to the first question, we have already seen that the available information restricts the learning scenario to supervised, reinforcement or unsupervised learning. We have also seen that the selection of relevant information is at the core of the credit assignment problem, and that the various learning rules that have been investigated in the different learning scenarios give partial answers to this question.

Common to all of the learning models described above is that learning is represented as determining a mapping f that transforms some input x into some suitable output y. The mapping f is constrained by several factors. First, it must belong to some pre-specified family F of admissible functions. Examples include the perceptron where F is the set of linear threshold functions, superpositions of radial basis functions, or more complex families, such as the transfer functions that result when the transformations of several perceptron layers are concatenated. In all these examples (but not necessarily in general), the family F is given as a set of parametrized functions. Learning can then be viewed as determining a particular member f of the family F which satisfies a number of further requirements that result from the posed learning task. As we have seen above, these requirements can be the minimization of a suitable cost function. However, the present formulation is more general and permits requirements for which the specification of a cost function may not be obvious. Usually, F is a very large set, and there may be a finite or infinite number of members f that satisfy all requirements of the learning task. On the other hand, if no member of F satisfies the requirements of the task, the learning task may be unsolvable. Still, F may contain one or many functions that can approximately satisfy the requirements of the task, and that may, therefore, yield useful approximations. In this case one can always reformulate and weaken the task requirements such that these functions qualify as valid, but maybe suboptimal, solutions.

This raises the question of how to find good encodings that facilitate learning. One can list a number of properties that characterize good encodings. Generally speaking, good encodings support generalization, that is, situations that require similar actions are encoded by similar patterns. They also must capture the "essential" variability in the data.

An important property required of network solutions f is the capability to generalize. f ought to perform well on previously unencountered samples of the task. Intuitively this means that the network ought to capture the essence of the task, and not just (over)fit the examples. Mathematically, it means that for a given network the number of learning samples is related to a particular generalization capability, that is, to the probability of committing an error over the whole sample space. One way to spell out this relation is provided by the theory of probably approximately correct learning (PAC-learning, Valiant 1984; Blumer et al. 1987). This theory considers, how the sample of m training examples restricts the space F of possible network instantiations such that for one of the remaining solutions f (a solution which is compatible with all the presented samples) the expectation value for an error on the whole sample set is low. The scheme involves two probabilities, one which reflects the effectiveness of a set of m randomly chosen learning samples for restricting the space F, and a second one, which reflects the probability of committing an error anyway. Clearly, a network with more weights has a larger space F of possible instantiations, and thus requires more examples m to be restricted.

A related concept is that of the Vapnik-Chervonenkis dimension (VC-dimension). For a given type of network, the VC-dimension is the maximum number of samples such that any Boolean function of these samples can be implemented with the network. Using the VC-dimension, one can also derive a bound on the required number of training samples, such that the resulting generalization error lies with some (high) probability below some (low) bound (Vapnik 1982; Haussler 1988). Clearly, both methods relate to one of the most central aspects of prerational intelligence, generalization capability. Yet, the safety of a rigorous mathematical result is bought at the expense of large requirements on the size of training sets. For perceptrons with W weights one ought to train with more than W/ϵ training samples in order to achieve a probability of $1 - \epsilon$ to classify any pattern of the pattern set correctly. A similar result holds for multilayer perceptrons, which also require more than 10 times as many training patterns as weights in the net in order to achieve a performance of 90% on the whole sample (Baum & Haussler 1989).

Finally we mention the problem of learnability. It is well known that simple perceptrons cannot learn linearly inseparable problems (Minsky 1988). On the other hand, multilayer-perceptrons with a single hidden layer can approximate functions to an arbitrary degree of accuracy (Funahashi 1989; Hornik et al. 1989). However, this and similar theorems give no bounds on the number of required neurons and, therefore, are mainly of theoretical interest.

A related problem is the "loading problem": Given a network architecture, characterized by some size parameter N (such as, e.g., the number of inputs to the network) and a function that is known to be implementable on this architecture by a suitable weight pattern, how difficult is the task of finding the weight values from the given data? It has been shown that functions can be constructed that make this task NP-hard, even if the number of layers and neurons of the network is fixed and very small (Judd 1988). However, it should be noted that this result rests on the choice of functions that are constructed to be "maximally unfavorable" for the network. Empirical work, employing improved learning strategies, such as learning by queries, has shown the practical solvability of the loading problem for networks with several hundred neurons (Baum 1991).

*Universität Bielefeld, Germany
**Universität Frankfurt, Germany

REFERENCES

Amit, D.J. (1989). *Modeling brain function: The world of attractor neural networks.* Cambridge, UK: Cambridge University Press.

Amit, D.H., H. Gutfreund, & H. Sompolinsky (1985). Storing infinite numbers of patterns in a spin-glass model of neural networks. *Phys. Rev. Lett.* **55**, 1530–1533.

Atick, J.J., & A.N. Redlich (1990). Towards a theory of early visual processing. *Neural Computation* **2** 308-320.

Atick, J.J., & A.N. Redlich (1992). What does the retina know about natural scenes? *Neural Computation* **4**, 196–210.

Babloyantz, A., & A. Destexhe (1986) Low-dimensional chaos in an instance of epilepsy. *Proc. Natl. Acad. Sci. U.S.A.* **83**, 3513–3517.

Baldi, P., & W. Heiligenberg (1988). How sensory maps could enhance resolution through ordered arrangements of broadly tuned receivers. *Biological Cybernetics* **59**, 313–318.

Barlow, H.B., & P. Földiák (1989). Adaptation and decorrelation in the cortex. In R. Durbin, C. Miall, & G. Mitchison (eds.), The computing neuron (pp. 54–72). Redwood City, CA: Addison Wesley.

Barlow, H.B. (1972). Single units and sensation: A neuron doctrine for perceptual psychology? *Perception* **1**, 371–394.

Basar, E. (ed.), (1990). *Chaos in brain function.* Series in Brain Dynamics. Berlin: Springer.

Baum, E.B. (1991). Neural net algorithms that learn in polynomial time from examples and queries. *IEEE Transactions on Neural Networks* **2**, 5–19.

Baum, E.B., & D. Haussler (1989). What size net gives valid generalization? *Neuronal Computation* **1**, 151–160.

Becker, S., & G.E. Hinton (1992). Self-organizing neural network that discovers surfaces in random dot stereograms. *Nature* **355**, 161–163.

Bergman, A., P. Grassberger, & T.P. Meyer (1991). *Forecasting probabilities with neural networks.* Preprint WU-B-91-8, University of Wuppertal, Germany.

Bialek, W. (1992). *Optimal signal processing in the nervous system.* Princeton Lectures on Biophysics. Singapore: World Scientific Publishing.

Bialek, W., F. Rieke, R.R. de Ruyter van Steveninck, & D.Warland (1991). Reading a neural code. *Science* **252**,1854–1857.

Blumer, A., A. Ehrenfeucht, D. Haussler, & M.K. Warmuth (1987). Occam's razor. *Information. Processing Letters* **24**, 377–380.

Churchland, P.S., & T.J. Sejnowkski (1993). *The computational brain.* Cambridge, MA: MIT Press.

Cohen, M.A., & S. Grossberg (1983). Absolute stability of global pattern formation and parallel memory storage by competitive neural networks. *IEEE Transactions on Systems, Man and Cybernetics* **13**, 815–826.

Davis, D.D., & C.A. Holt (1992). *Experimental economics.* Princeton, NJ: Princeton University Press.

Daugman, J.G. (1985). Uncertainty relation for resolution in space, spatial frequency, and orientation optimized by two-dimensional visual cortical filters. *Journal Optical Society America* **A 2**, 1160–1169.

de Sa, V. (2000). Combining uni-modal classifiers to improve learning. In H. Ritter, H. Cruse, & J. Dean (eds.), *Prerational intelligence: Adaptive behavior and intelligent systems without symbols and logic,* Vol. 2 (pp. 707–722). Dordrecht, The Netherlands: Kluwer Academic Publishers.

Debenham, R.M., & S.C.J. Garth (1989). Investigations into the effect of numerical resolution on the performance of back propagation. In L. Personnaz & G. Dreyfus (eds.), *Neural networks from models to applications* (pp. 752–755). Paris: IDSET.

Degn, H., A.V. Holden, & L.F. Olsen (eds.), (1987). Chaos in biological systems. *NATO ASI Series A (Life Sciences),* Vol. **138**.

Dijkstra, T., G. Schöner, & C. Gielen (1994). Temporal stability of the action-perception cycle for postural control in a moving visual environment. *Biological Cybernetics* **97**, 477–486.

Duda, R.D., & P.E. Hart (1973). *Pattern classification and scene analysis.* New York: John Wiley.

Durbin, R., & D.J. Willshaw (1987). An analogue approach to the traveling salesman problem using an elastic net method. *Nature* **326**, 689–691.

Engel, A.K., P. König, A.K. Kreiter, & W. Singer (1991). Interhemispheric synchronization of oscillatory neuronal responses in cat visual cortex. *Science* **252**, 1177–1179.

Enquist, M., & A. Arak (1994). Symmetry, beauty and evolution. *Nature* **372**, 169–172.

Erwin, E., K. Obermayer, & K. Schulten (1992). Self-organizing maps: Ordering, convergence properties and energy functions. *Biological Cybernetics* **67**, 47–55.

Fahlman, S.E. (1988). Fast-learning variations on back-propagation: An empirical study. In D. Touretzky, G. Hinton, & T. Sejnowski (eds.), *Proceedings of the 1988 Connectionist Model Summer School* (Pittsburgh) (pp. 38–51). San Mateo, CA: Morgan Kaufman.

Fitzhugh, R. (1961). Impulses and physiological states in theoretical models of nerve membranes. *Biophysics Journal* **1**, 445–446.

Freeman, W.J. (1991). The physiology of perception. *Scientific American* (February 1991), 34–41.

Funahashi, K.-I. (1989). On the approximate realization of continuous mappings by neural networks. *Neural Networks* **2**, 183–192.

Gabor, D. (1946). Theory of communication. *Journal of the Institute of Electrical Engineering* **93**, 429–457.

Gardner, E. (1987). Maximum storage capacity in neural networks. *Europhysics Letters* **4**, 481–485.

Gat I., & N. Tishby (1993). Statistical modeling of cell-assemblies activities in associative cortex of behaving monkeys. In J. Hanson, J.D. Cowan, & C.L. Giles (eds.), *Advances in neural information processing systems,* 5 (pp. 945–952). San Mateo, CA: Morgan Kaufmann.

Gierer, A., & H. Meinhard (1972). A theory of biological pattern pormation. *Kybernetik* **12** 30–39.

Glass, L., & D. Kaplan (1994). Complex dynamics in physiology and medicine. In A.S. Weigend & N.A. Gershenfeld (eds.), *Time series prediction* (pp. 513–528). Santa Fe Institute Proceedings Vol. 15.

Goldberg, D.E. (1989). *Genetic algorithms in search, optimization and machine learning.* Redwood City, CA: Addison Wesley.

Grassberger, P., & I. Procaccia (1983). Characterization of strange attractors. *Physical Review Letters* **50**, 346–349.

Hansel, D., & H. Sompolinsky (1992). Synchronization and computation in a chaotic network. *Physical Review Letters* **68**, 718–721.

Haussler, D. (1988). Quantifying inductive bias: AI learning algorithms and Valiant's learning framework. *Artificial Intelligence* **36**, 177–222.

Heiligenberg, W. (2000). The Jamming Avoidance Response (JAR) of the electric fish *Eigenmannia*: The processing of sensory information and motor control. In H. Cruse, H. Ritter, & J. Dean (eds.), *Prerational intelligence: Adaptive behavior and intelligent systems without symbols and logic,* Vol. 1 (pp. 59–83). Dordrecht, The Netherlands: Kluwer Academic Publishers.

Hinton, G.E., & T.J. Sejnowski (1986). Learning and relearning in Boltzmann machines. In D.E. Rumelhart & J.L. McClelland (eds.), *Parallel distributed processing,* Vol. 1 (pp. 282–317). Cambridge, MA: MIT Press.

Holland, J.H. (1975). *Adaptation in natural and artificial systems.* Ann Arbor, MI: University of Michigan Press.

Hopfield, J.J. (1982). Neural networks and physical systems with emergent collective computational abilities. *Proceedings of the National Academy of Sciences, USA,* **79**, 2554–2558.

Hopfield, J.J. (1984). Neurons with graded response have collective computational properties like those of two-state neurons. *Proceedings of the National Academy of Sciences, USA,* **81**, 3088–3092.

Hornik, K., M. Stinchcombe, & H. White (1989). Multilayer feedward networks are universal approximators. *Neural Networks* **2**, 359–366.

Huberman, B.A., & N.S. Glance (1993). Evolutionary games and computer simulations. *Proceedings of the National Academy of Sciences, USA,* **90**, 7716–7718.

Johnstone, R. (1994). Female preference for symmetrical males as a by-product of selection for mate recognition. *Nature* **372**, 172–175.

Judd, S. (1988). On the complexity of loading shallow networks. *Journal of Complexity* **4**, 177–192.

Julesz, B. (1981). Textons, the elements of texture perception, and their interactions. *Nature* **290**, 91–97.

Kaas, J.H., M.M. Merzenich, & H.P. Killackey (1983). The reorganization of somatosensory cortex following peripheral nerve damage in adult and developing mammals. *Annual Review of Neuroscience* **6**, 325–356.

Kappen, H.J. (2000). Constructing modular architectures for coordination in machines past and present. In H. Ritter, H. Cruse, & J. Dean (eds.), *Prerational intelligence: Adaptive behavior and intelligent systems without symbols and logic,* Vol. 2 (pp.157–164). Dordrecht, The Netherlands: Kluwer Academic Publisher.

Kirkpatrick, S., C.G. Gelatt Jr., & M.P. Vecchi (1983). Optimization by simulated annealing. *Science* **220**, 671–680.

Koenderink, J. (1989). *Solid shape.* Cambridge, MA: MIT Press

Kohonen, T. (1982). Self-organized formation of topologically correct feature maps. *Biological Cybernetics* **43**, 59–69.

Kohonen, T. (1989). *Self-organization and associative memory.* Heidelberg, Germany: Springer-Verlag.

Kreiter, A.K., & W. Singer (1992). Oscillatory neuronal responses in the visual cortex of the awake Macaque monkey. *European Journal of Neuroscience* **4**, 369.

Krüger, J., & J.D. Beckers (1991). Recognizing the visual stimulus from neuronal discharges. *Trends in Neurosciences* **14**, 282–286.

Kuramoto, Y. (1991). Collective synchronization of pulse-coupled oscillators and excitable units. *Physica* **D 50**, 15–30.

Linsker, R. (1986). From basic network principles to neural architecture. *Proceedings of the National Academy of Sciences, USA,* **83**, 7508–7512, 8390–8394, 8779–8783.

Luttrell S.P. (1989). Self-organisation: A derivation from first principles of a class of learning algorithms. *Proc. 3rd. IEEE Int. Joint Conf. Neural Networks*, Washington DC **2**, 495–498.

Malsburg von der, C. (1973). Self-organization of orientation sensitive cells in the striate cortex. *Kybernetik* **14**, 85–100.

Malsburg von der, C. (1981). *The correlation theory of brain function.* Internal Report 81-2, Max-Planck-Institute for Biophysical Chemistry, Göttingen, Germany.

Malsburg von der, C. (1986). Am I thinking assemblies? In G. Palm & A. Aertsen (eds.), *Brain theory* (pp. 161–176). Heidelberg-Berlin: Springer-Verlag.

Martinetz, T., H. Ritter, & K. Schulten (1988). Kohonen's self-organizing map for the modeling of the auditory cortex of a bat. In *SGAICO-Proceedings "Connectionism in Perspective"* (pp. 403–412). Zürich, Switzerland.

Metropolis, N., A.W. Rosenbluth, M.N. Rosenbluth, A.H. Teller, & E. Teller (1953). Equation of state calculations for fast computing machines. *Journal of Chemical Physics* **21**, 1087–1092.

Milner, P.M. (1974). A model for visual shape recognition. *Psychological Review* **81**, 521–535.

Minsky, M.L., & S.A. Papert (1988). *Perceptrons* (expanded ed.). Cambridge, MA: MIT Press.

Newsome, W.T., K.H. Britten, & J.A. Movshon (1989). Neuronal correlates of a perceptual decision. *Nature* **341**, 52–54.

Nowak, M.A., & R.M. May (1992). Evolutionary games and spatial chaos. *Nature* **359**, 826–829.

Obermayer, K. (1990). A principle for the formation of the spatial structure of cortical feature maps. *Proceedings of the National Academy of Sciences, USA*, **87**, 8345–8349.

Oja, E. (1982). A simplified neuron model as a principal component analyzer. *Journal of Mathematical Biology* **15**, 267–273.

Optican, L.M., & B.J. Richmond (1987). Temporal encoding of two-dimensional patterns by single units in primate inferior temporal cortex. III. Information theoretic analysis. *Journal of Neurophysiology* **57**, 162–178.

Ott, E., C. Grebogi, & J.A. Yorke (1990). Controlling chaos. *Physical Review Letters* **64**, 1196–1199.

Pawelzik, K., H.-U. Bauer, & T. Geisel (1993). Alternating predictable and unpredictable states in data from cat visual cortex. In F.H. Eeckman & J. Bower (eds.), *Computation and neural systems* (pp. 487–494). Boston, MA: Kluwer Academic Publishers.

Poggio, T., V. Torre, & C. Koch (1985). Computational vision and regularization theory. *Nature* **317**, 314–319.

Poggio, T., & T. Vetter (1992). *Recognition and structure from one 2d model view: Observations on prototypes, object classes and symmetries*. AI Memo 1347. Cambridge: MIT AI Laboratory.

Radons, G., J.D. Beckers, B. Dülfer, & J. Krüger (1994). Analysis, classification and coding of multielectrode spike trains with Hidden Markov Models. *Biological Cybernetics* **71**, 359–373.

Rechenberg, H. (1973). *Evolutionsstrategie – Optimierung technischer Systeme nach den Prinzipien der biologischen Evolution*. Stuttgart, Germany, Frommann-Holzboog Verlag.

Richmond, B.J., & L.M. Optican (1990). Temporal encoding of two-dimensional patterns in primate primary visual cortex. II. Information transmission. *Journal of Neurophysics* **64**, 370–380.

Ritter, H., & K. Schulten (1986). On the stationary state of Kohonen's self-organizing sensory mapping. *Biological Cybernetics* **54**, 99–106.

Ritter, H., & T. Kohonen (1989). Self-organizing semantic maps. *Biological Cybernetics* **61**, 241–254.

Rolls, E. (1989). The representation and storage of information in neural networks in the primate cerebral cortex and hippocampus. In R. Durbin, C. Miall, & G. Mitchison (eds.), *The computing neuron* (pp. 125–159). Redwood City, CA: Addison Wesley.

Rubner, J., & P. Tavan (1989). A self-organizing network for principal-component analysis. *Europhysics Letters* **10**, 693–698.

Rumelhart, D.E., G.E. Hinton, & R.J. Williams (1986). Learning representations by back-propagating errors. *Nature* **323**, 533–536.

Sagerer, G., & H. Niemann (1997). Semantic networks for understanding scenes New York : Plenum. (Advances in computer vision and machine intelligence)

Schöner, G. (2000). What can we learn from dynamic models of rhythmic behavior in animals and humans? In H. Ritter, H. Cruse, & J. Dean (eds.), *Prerational intelligence: Adaptive behavior and intelligent systems without symbols and logic,* Vol. 2 (pp. 223–241). Dordrecht, The Netherlands: Kluwer Academic Publishers.

Schöner, G., & J.A.S. Kelso (1988). Dynamic pattern generation in behavioral and neural systems. *Science* **239**, 1513–1520.

Schmidhuber, J. (2000).Neural predictors for detecting and removing redundant information. In H. Ritter, H. Cruse, & J. Dean (eds.), *Prerational intelligence: Adaptive behavior and intelligent systems without symbols and logic,* Vol. 2 (pp. 691–706). Dordrecht, The Netherlands: Kluwer Academic Publishers.

Schuster, H.G. (1989). *Deterministic chaos.* (2nd ed.) Weinheim, Germany: Verlag Chemie.

Sejnowski, T.J., & C.R. Rosenberg (1987). Parallel networks that learn to pronounce English text. *Complex Systems* **1**, 145–168.

Selten, R. (ed.), (1991). *Game equilibrium models I-IV.* Heidelberg-Berlin: Springer-Verlag.

Skarda, C., & W.J. Freeman (1987). How brains make chaos in order to make sense of the world. *Behavioral Brain Sciences* **10**, 161–195.

Smagt, P. van der (1994). Minimisation methods for training feedforward neural networks. *Neural Networks* **7**, 1–11.

Smithers, T. (2000). On behaviour as dissipative structures in agent-environment interaction systems. In H. Ritter, H. Cruse, & J. Dean (eds.), *Prerational intelligence: Adaptive behavior and intelligent systems without symbols and logic,* Vol. 2 (pp. 243–257). Dordrecht, The Netherlands: Kluwer Academic Publishers.

Solla, S.A., E. Levin, & M. Fleisher (1988). Accelerated learning in layered neural networks. *Complex Systems* **2**, 625–639.

Sompolinsky, H., D. Golomb, & D. Kleinfeld (1991). Cooperative dynamics in visual processing. *Physical Review Letters* **43**, 6990–7011.

Strogatz, S.H. (1994). *Nonlinear dynamics and chaos.* Reading, PA: Addison Wesley.

Takens, F. (1980). Detecting strange attractors in turbulence. In D.A. Rand & L.S. Young (eds.), *Dynamical systems and turbulence* (pp. 336–381). Springer Lecture Notes in Mathematics 898. Berlin: Springer-Verlag.

Thorndyke, E.L. (1966). *Human learning*. Cambridge, MA: MIT Press.

Tsodyks, M.V., & M.V. Feigel'man (1988). The enhanced storage capacity in neural networks with low activity level. *Europhysical Letters* 6, 101–105.

Turing, A.M. (1952). The chemical basis of morphogenesis. *Philosophical Transactions of the Royal Society London B* 237 37–72.

Valiant, L.G. (1984). A theory of the learnable. *Communications of the Association for Computing Machinery* 27, 1134–1142.

Vapnik, V.N. (1982). *Estimation of dependencies based on empirical data*. Berlin: Springer-Verlag.

Vetter T., T. Poggio, & H.H. Bülthoff (1994). The importance of symmetry and virtual views in 3-dimensional object recognition. *Current Biology* 4, 18–23).

Watkins, C.J.C.H. (1989). *Learning from delayed rewards*. PhD thesis, Cambridge, UK.

Wehrhahn, C. (2000) Retinal coding of Vernier acuity and motion. In H. Cruse, H. Ritter, & J. Dean (eds.), *Prerational intelligence: Adaptive behavior and intelligent systems without symbols and logic,* Vol. 1 (pp. 151–162). Dordrecht, The Netherlands: Kluwer Academic Publishers.

Westheimer, G., & S.P. McKee (1977). Spatial configurations for hyperacuity. *Visual Research* 17, 941–947.

Willshaw, D.J., & C. von der Malsburg (1976). How patterned neural connections can be set up by self-organization. *Proceedings of the Royal Society of London B* 194, 431–445.

Wilson, H.R., & J.D. Cowan (1972). Excitatory and inhibitory interactions in localized populations of model neurons. *Biophysical Journal* 12, 1–24.

Winfree, A.T. (1987). *When time breaks down: The three-dimensional dynamics of chemical waves and cardiac arhythmics*. Princeton, NJ: Princeton University Press.

Wolf F., H.-U. Bauer, & T. Geisel (1994). Formation of field discontinuities and islands in visual cortical maps. *Biological Cybernetics* 70, 525–531.

Wolf, F., K. Pawelzik, T. Geisel, D.S. Kim, & T. Bonhoeffer (1994). Optimal smoothness of orientation preference maps. In F.H. Eeckman (ed.), *Computation in neurons and neural systems* (pp. 103–108). Dordrecht, The Netherlands: Kluwer Academic Publishers.

PETER LANZ* and DAVID McFARLAND**

PHILOSOPHICAL PERSPECTIVES ON REPRESENTATION, GOALS, AND COGNITION

1. INTRODUCTION

In this paper we address three concepts that are much talked about in the animal robotics community. These concepts are (1) representations, (2) goals, (3) minimal cognition.

We want to distinguish between information as an objective commodity and representation as something which involves a user, i.e., a system which accesses and uses information. Information per se lies out there and exists independently of any system that makes use of it. Representations presuppose design and require a user.

We want to distinguish different kinds of purposive behavior, which involve different notions of goals, which themselves appear to be different from the point of view of the agent, the designer and the naive observer.

We want to try to define cognition in terms of the minimal criteria that would be acceptable in view of our definitions of representations and goals. We recognise that full blown cognition may require ingredients other than those of minimal cognition.

2. REPRESENTATIONS

2.1 *Information as an Objective Commodity*

We begin by explaining the notion of information. All effects bear information about their causes; a geologist can look at stones for the information they contain about the early history of the earth. Tree rings contain information about the age of the tree. In some sense they (naturally) represent the age of the tree. Footprints on the sand bear information about the nearby presence of animals. Smoke contains information about fire, etc. The information is out there. Therefore, information is an objective commodity. In itself this information is inert. Trees do not make any use of the fact that their rings indicate their age. To make a difference to the course of events, information has to be accessed somehow, to be picked up and used. This sets up demands for users. A prospective user has to develop or somehow to acquire devices for picking up information and for putting it to some use. This

247

J. Dean et al. (eds.), *Prerational Intelligence: Interdisciplinary Perspectives on the Behavior of Natural and Artificial Systems*, 247–263.
© 2000 *Kluwer Academic Publishers*.

requires a corresponding design (implicit representation) or the capacity to construct and use explicit representations (see below).

Information is always veridical, because the condition for there being a certain state of affairs (say, fifty tree rings) coincides with the condition that this state of affairs bears information about the age of the tree (the tree is fifty years old). But what about the case in which the tree is fifty years old, but has only forty-nine tree rings due to the fact that there was no growth during a particular year? The fact that the tree is fifty years old is *not* the cause for there being forty-nine rings. Therefore, the forty-nine rings do not contain information about the exact age of the tree. It is a combination of the true age and the missing growth during one year which caused the forty-nine rings. Therefore, the forty-nine rings contain information about peculiarities of the growth of the tree. The forty-nine tree rings do not contain or bear misinformation about the age of the tree. Nevertheless, the forty-nine tree rings may prompt a human observer to *misrepresent* the true age of the tree, because he assumes that which is usually the case, namely, that the number of tree rings reliably indicates the age of the tree, because each year gives rise to one ring.

To take another example, normally, a car's petrol gauge indicates 'empty', when the tank is empty. It indicates the state which is, from the point of view of the designer, the prime causal factor of its own state. Therefore, the petrol gauge cannot (correctly) indicate, that the tank is empty, if the tank is not empty. If something goes wrong, then the pointer of the petrol gauge can point to the 'empty' sign without the tank being empty, in which case the petrol state is not said to be indicated. What the pointer on 'empty' sign indicates when the tank is not empty, is the fact that there is something wrong with the device. To say, that the petrol gauge 'misinforms' the customer only means, that it induces the costumer to misrepresent the state of the tank, because he assumes that all is well. We therefore conclude, that information is always objectively there, but the system that makes use of the information always has to rely on some assumptions about reliability, about the absence of 'counteracting causes'. In the case of artificial devices, it has to rely on assumptions about well-functioning, e.g., on the assumption that the device is performing as it is supposed to perform.

2.2 *On the Nature of Representation*

What is normally called a representation is something that can be semantically evaluated. A representation is said to have a content and to have satisfaction conditions, i.e., the state of affairs which has to obtain for the representation to be a true representation. A representation refers to or denotes something; it is about something and states something about that which it denotes. Representations are always partial or selective: they represent their objects always under one aspect or another. Representations can be evaluated as accurate, as reliable, as correct, as well-founded, etc.; representations are therefore also susceptible to epistemic evaluation.

There is one sense of the word 'representation' in which it can be used interchangeably with 'information'. We call this 'natural representation'. Natural representations, such as smoke for fire, footprints for animals, etc., do not mean more than "indication" or "registration". Artifactual representations, on the other hand, do have a different connotation. The following characteristics of artefactual representations are important: *(i)* There has to be some token or symbol of the representational system (e.g. traffic signs, Chinese characters, etc.) (Harnard 1990, p. 336). *(ii)* The token or symbol is supposed to have a certain meaning. *(iii)* There has to be a potential user of the tokens of the representational system. *(iv)* The use of the token should not be fortuitous. *(v)* The tokens can not only be used to represent something, but also to misrepresent the same thing (e.g. a school sign may remain, even though the school is no longer functioning).

One may put the distinction between information and artefactual representation in more formal terms:

(a) R represents x by virtue of covarying with x (of being reliably correlated with x; of indicating x, etc.). Causal or probabilistic relations ground this "covarying" and "being correlated with". This captures "indication" and "registration". This formulation underlines the fact, that events contain information about other events as a matter of the holding of natural laws.

(b) R represents x by virtue of its *functional role* in the system S. This functional role normally involves the following two things: (1) Occurrences of R help the system to keep track of certain environmental features. We call this the etiological aspect of the functional role of R. (2) The fact, that Rs reliably track certain environmental features, helps the system to orient its own behavior towards these features in beneficial ways, given its own needs. We call this the forward-looking aspect of the functional role of R. The main point here is the following: The representation of a feature (or the knowledge of the presence of that feature) helps the system to correlate the fact, that a certain behavior would be beneficial in these circumstances, with the ability of the system to manifest this behavior under the circumstances. To call this 'artefactual' representation does not mean that this type of representation does not occur in nature; of course, it does. The word 'artefactual' is used to point out the fact, that representation, in contrast to information, requires some design. The design may be a result of evolution (natural selection), of learning or training or of the decisions of engineers. Without design, whatever its source or origin, there would be no supposition about what a token or symbol means and therefore, there would be no possibility for misrepresentation. Note that something which is not supposed to have any function can also not fail to function correctly.

We conclude that the use of the word 'representation' is always appropriate when there is information (or misrepresentation of information) as a function of design and not only as a function of the fact that events fall under natural laws.

2.3 *Implicit and Explicit Representations*

Not every representation has to be an explicit representation. We propose the following: Implicit representations are the same as information that is available for use by a competent user. For example, a diskette displays information about top-underside, which is important if the user is to insert the diskette the correct way up. If the designer of the diskette is satisfied that every user will recognise the topside by signs that are inevitably present on all diskettes, then he may decide to add no further information. If he decides otherwise, then he may add a symbol, or explicit representation of the topside. In the former case, the required information is already implicit in the design of the diskette, and we may say that 'topside' is implicitly represented. In the latter case, topside is explicitly represented, because the designer decided that this was necessary. Therefore, we conclude that implicit representations are equivalent to information that is available to a competent user.

It is a form of representation, because the state of affairs may be misrepresented. For example, it may be that the diskette reader is installed in the computer in an unconventional orientation, so that the normal topside is now not the side that should be on top when the diskette is inserted. Compare also the difference between a system which merely executes the instruction "When A, then B" and a system which consults the rule "When A, then B". In the former case the instruction is implicitly represented; in the second case the rule is explicitly represented. This difference is also expected to manifest itself in the behavior of a system (natural or artificial), for example in the type of errors to which the respective systems are liable.

Representations can be implicit in more than one way. (i) The information is somehow there, but has to be inferred. One can infer from the structure of a diskette that one side is the top side. (ii) Implicit can also mean 'embedded' (McFarland & Bösser 1993). For example, body temperature is measured by sensors which are calibrated so that the outcome is in tune with other bodily processes. Implicit in the calibrated information is a transfer function or conversion process that is not cognitively appreciated by the agent, but is nevertheless embedded in the structure. It is useful to use the term implicit for cases where the correspondence is mathematically implicit, and the term embedded where it is structurally implicit.

Consider the concept of the "implicit world model" (e.g., Shepard 1987). There is in vision research the following discovery: a vision module designed to extract a 3-D shape from stereoscopic images works rapidly if it is equipped with an algorithm which differentiates. But such an algorithm only works well when the following assumption about the world is true: objects change in shape smoothly and continuously. Now, in most cases this assumption is true. This does not mean that the system has an implicit theory of the world. That the assumption is true is the designer's justification for choosing algorithms with certain properties. The assumption of smoothness nowhere is part of what the system itself states or represents or informs others about. Generalized: If a certain procedure achieves a certain end in a reliable way, then one can always discover features of the world on which the reliability of the procedure depends. If the world had had other properties, then other procedures would have been reliable. But does this justify us in saying the system has an implicit world model? Our answer is that this is yet another meaning of the word implicit. We have distinguished between mathematically and structurally (embedded) implicitness. Here the implicitness rests on a 'presupposition' by the designer. The presupposition is that the environment in which the agent is supposed to work has certain properties. We reserve the word 'theory' for the explicit statement of the relevant properties an environment is believed to have. In other words, when people use the phrase 'implicit theory' they should mean nothing more than this: a particular system works well in a certain environment, because the designer of the system (nature or engineer) did take into account the fact that this environment has these and those properties. The fact that muscles and tendons of animals are well adapted to the effects of gravity should not invite us to say, that muscles and tendons have an implicit theory of gravity.

We move now to explicit representations. By explicit we mean that the information is made obvious in a physical manner, and is not simply part of a procedure. If a representation is to be explicit, then there has to be a physically identifiable bearer of the information (the token) and, additionally, something which can be identified as the user of the representation.

The notion of explicit representation is not independent of the specification of how a system uses the information. Explicitness is not a local property, but a systemic property. One has to take into account both, representation and its use by a system.

That part of the system that makes use of the information, we will call the 'user'. A system involving explicit representations must be defined in such a way as to include the user. For example, many motor cars have icons on the dash which light up when there is a certain kind of fault. Such icons provide information, but if we are considering only the car as the system, they cannot be said to be explicit representations. If, however, we count both car and driver as the system, then the icons explicitly represent some fault in the system. In the case of implicit representations, there is no user. We will illustrate this point by reference to a specific robot example.

Holland et al. (1994) built battery-powered robots on a 21×17cm plat-form. A 12V motor powered wheel was positioned at the mid-point of each long side, with a castor wheel at the mid-point of one of the shorter sides; this allows the robot to move forwards or backwards in a straight or curved trajectory, and to turn on the spot. Each robot carries a 17cm wide alu-minium forward-facing C-shaped horizontal scoop with which it can push objects. The objects used are circular pucks, 4cm in diameter and 2.5cm in height. The robots are equipped with two IR sensors for obstacle avoidance, and a microswitch which is activated by the scoop when a certain number of pucks are pushed. For the experiments reported here, this number is set to three.

The robots have only three behaviours, and only one is active at any time. When no sensor is activated, a robot executes the default behavior of moving in a straight line until an obstacle is detected by the IR sensors, or until the microswitch is activated (pucks are not detected by the IR sensors). On detecting an obstacle, the robot executes the obstacle avoidance behavior of turning on the spot away from the obstacle and through a random angle; the default behavior then takes over again, and the robot moves in a straight line in the new direction. If the robot is pushing pucks when it encounters the obstacle, the pucks will be retained by the scoop throughout the turn. When the scoop pushes three or more pucks, the microswitch is activated; this triggers the puck-dropping behavior, which consists of backing up by reversing both motors for 1 second (releasing the pucks from the scoop), and then executing a turn through a random angle, after which the robot returns to its default behavior and moves forwards in a straight line. The obstacle avoidance behavior has priority over the puck-dropping behavior.

The robots operate completely autonomously and independently; all sen-sory, motor, and control circuitry is on board, and there is no explicit com-munication (IR or radio link) with other robots or with the experimenters. The robots only react to the local configuration of the environment.

The IR sensors have to be calibrated to be attuned to the environment in which the robots are operating. Such calibration constitutes 'embedded' information, a form of implicit representation (see above). Similarly, the microswitch is calibrated to trigger on three pucks, and it also provides information to the rest of the system. Both types of sensor can be said to provide implicit representations, of obstacles and pucks respectively, because both are tied to fixed procedures.

What would be required for these sensors to provide explicit representations of obstacles and pucks respectively? There would have to be a token supposed to represent an obstacle (or puck), and a system capable of taking the representation as representing an obstacle (or puck). What use the system makes of the representation is a matter for the designer. For example the designer of a computer diskette can provide a symbol indicating topside, but this is not necessary if all diskette readers are installed the same way up (see above). If however, some are installed the other way up, then the designer would be prudent to provide an explicit representation of the topside of the diskette and an explicit representation of the orientation of the reader in the computer.

In the case of the robot, the use made of information about obstacle (or pucks) is always the same. Therefore it is not necessary to provide an explicit representational system, although this could be provided as a luxury.

In summary, implicit representations make information available for use, but do not require a user. Explicit representations require a token, an interpreter, and a user of the representational system.

3. GOALS

There are three types of apparently purposive behavior, that we wish to discuss.

1. A goal-achieving system is one which can recognise the goal once it is arrived at (or at least change its behavior when it reaches the goal), but the process of arriving at the goal is largely determined by the environmental circumstances (McFarland & Bösser 1993).

To recognise a goal an explicit representation of the state of affairs that constitutes the goal would be necessary (because recognition implies cognition). Goal-achieving behavior can also be achieved by a system that has only an implicit representation of the goal, provided the received information is instrumental in changing the behavior of the agent. For example, the robots (discussed above) are equipped with a microswitch which is activated by the scoop when three pucks are pushed. The microswitch triggers the puck-dropping behavior, which consists of backing up, releasing the pucks from the scoop, and then executing a turn through a random angle, after which the robot returns to other behavior. The microswitch provides a representation of the pucks, which is implicit because the designer has not provided an explicit representational system. This is because the use made of the information is entirely procedural, so no explicit representational system is necessary. The received information is instrumental in changing the behavior of the robot, so this condition for goal-achieving is satisfied.

The second condition for defining a system as goal-achieving is the absence of processes within the agent that guide the agent towards the goal. In our robot example, there are no such processes. The robot arrives in a situation of having three pucks in its scoop in an entirely haphazard manner. Moreover the robot has no mechanisms for guiding its behavior. It switches among its three possible activities on the basis of encountering environmental situations that are outside its control.

From the point of view of the designer, there can be little question as to whether a given agent is goal-achieving. The designer has simply to identify the 'recognition' element of the system, and to confirm that there are no mechanisms within the agent that guide the agent towards the goal.

From the point of view of an observer, the view that the agent is goal-achieving is an hypothesis concerning the two conditions that are necessary for goal-achieving behavior. There must be a hypothesis about the use of information (i.e., 'recognising'), and af hypothesis about whether or not the behavior is internally steered towards the goal.

Bertrand Russell (1921) put forward a goal-achieving type of theory as a general theory of purposive behavior: "A hungry animal is restless until it finds food: then it becomes quiescent. The thing which will bring a restless condition to an end is said to be what is desired (its purpose)" (p. 32). The implication here is that the animal is in an environment in which the restless behavior is appropriate. When it encounters, and recognises, the relevant (food) stimuli, the animal changes its behavior.

To confirm the hypothesis that the behavior was goal-achieving, it would have to be shown that the 'end-state' is registered by the agent, and that the 'restlessness' in no way steers the agent towards this end-state.

Systems are goal-achieving by virtue of the fact that matters are so arranged that the necessary environmental features are generally present at the appropriate stage in the causal chain.

Goal-achieving behavior in humans is probably more commonplace than we tend to imagine. Consider the collector of matchboxes, or any other artefact with a long and varied history. The collector does not deliberately set out to search for matchboxes, but relies upon serendipity – the habit of making happy and unexpected discoveries by accident. The main characteristic of goal-achieving behavior is preprogrammed recognition. The goal is achieved by being in the right place at the right time, and recognising this state of affairs.

2. A goal-seeking system   is one which is designed to seek a goal without the goal being represented within the system. Many physical systems are goal-seeking in this sense. For example, a marble rolling around a bowl will always come to rest in the same place. It may take various different routes, depending upon the starting conditions. The marble appears to be goal-seeking, because the forces acting on it are so arranged that the marble 'is pulled' towards the goal.

From the point of view of the agent, the goal does not feature, because the agent has no information about the goal. Take, for example the experiments with robots by Holland et al. (1994). From a qualitative point of view, each experiment has three more or less distinct phases, regardless of the number of robots. At the start, the arena contains only single pucks. In the first phase, a robot typically moves forwards collecting pucks into the scoop one at a time. When three have been gathered, the robot drops them, leaving them as a cluster of three, and moves off in another direction. Within a short time, most pucks are in small clusters which cannot be pushed around. In the second phase, the robot removes one or two pucks from clusters by striking the clusters at an angle with the scoop. The pucks removed in this way are added to other clusters when the robot collides with them. Some clusters grow rapidly in this phase. After a time, there will be a small number of relatively large clusters. The third and most protracted phase consists of the occasional removal of a puck or two from one of the large clusters, and the addition of these pucks to one of the clusters, often to the one they were taken from in the first place. The process eventually results in the formation of a single cluster. The robot has no information about the existence of this single cluster. It cannot distinguish between a cluster of three pucks and a cluster of 30 pucks.

From the point of view of the designer, the system is goal-seeking by virtue of the fact that the designer provides both the agent and its environment. The designer provides both the marble and the bowl. The robot designer provides the robot and its test-environment. In the case of natural systems, the designer (evolution) provides an agent that is supposed to live in a particular environment.

From the point of view of the observer, the behavior of a goal-seeking agent appears to be purposive because the observer tends to identify certain types of agent-environmental configurations as goals (McFarland 1989). In the robot example, it seems obvious to the naive observer that the 'goal' of the robots is to collect all the pucks into a single cluster. Moreover, the agent appears to be persistent with respect to the 'goal'. In other words, there is a tendency to ascribe goals to situations that result from persistent behavior.

From the point of view of the agent, the goal-seeking system achieves its effects by virtue of the forces acting on and within the system. There is no internal representation of the goal-to-be-achieved.

4. A Goal-Directed System

A goal-directed system involves a representation of the goal-to-be-achieved, which is instrumental in directing the behavior. We reserve the term goal-directed to indicate behavior (of a human, animal, or machine) that is directed by reference to an internal (explicit or implicit) representation of the goal-to-be-achieved. By directed we mean that the behavior is actively controlled by reference to the (internally represented) goal. The behavior will be subject to outside disturbances, which will usually be corrected for. Thus by directed we mean that the behavior is guided or steered towards the goal, despite disturbances.

What would it take for the robots to be goal directed? The robot would have to have a representation of the state of affairs that would pertain if the goal were to be achieved – in other words, a representation of the pucks in a cluster. It might be possible to engineer such a representation simply by providing the robot with a second microswitch, with a stronger spring, which triggered at a force representative of the whole cluster (statistically the number of pucks in line with the scoop is related to the size of the cluster). Such a microswitch would provide an implicit representation of the cluster, in the case where the use made of the information is purely procedural, and an explicit representation in the case where the information could be put to various uses.

In addition, the representation of the goal-to-be-achieved would have to be instrumental in guiding the behavior. The second microswitch would not be adequate to satisfy this condition, because it gives information only about the end-state. Such information could not serve to guide the behavior prior to the end-state being achieved. To provide information adequate to guide the behavior the cluster-sensor would have to provide continuous information about the current size of the cluster.

From the point of view of the agent, the achievement of a goal can be deduced from comparison of the representation of the goal-to-be-achieved and the actual state of affairs. When this 'error' is zero the goal has been achieved.

From the point of view of the designer, it is clear when the goal has been achieved, because the designer knows with what representation of the goal the agent is equipped. Moreover, the designer can see the external affairs and can see the match between goal and representation. The designer can also see whether the agent's behavior conforms with expectations.

From the point of view of the naive observer, it is virtually impossible to tell when the agent has reached its goal. The observer has no information about the agent's representations, and can only form hypotheses based upon observation of the behavior.

To recapitulate, a system can be goal-achieving or goal-seeking without being goal-directed. A goal-achieving system is one which can recognise the goal once it is arrived at (or at least change its behavior when it reaches the goal), but the process of arriving at the goal is largely determined by the environmental circumstances. A goal-seeking system is one which is designed to seek the goal without the goal being explicitly represented within the system. Goal-seeking systems can be based upon a dynamic equilibrium of various forces operating within the system. A goal-directed system involves an explicit or implicit representation of the goal-to-be-achieved, which is instrumental in directing the behavior.

The distinctions among these different types of purposive behavior can also be made by means of the frame of reference. In the case of goal-achieving behavior, the designer can identify the goal, the agent can change its behavior when it reaches a goal, but the naive observer can only form hypotheses about the goal. In the case of goal-seeking behavior, the agent has no information about the goal. The designer can identify the goal of the agent-environment system. The naive observer is inclined to jump to conclusions about the goal, but has no way of knowing. In the case of a goal-directed system, the agent does have reliable information about the goal, as does the designer. The naive observer can only form hypotheses.

Take the case of an agent that exhibits apparently purposive behavior. The naive observer, or even the not so naive observer, is in a hopeless position. The agent may in fact be one of any of the three types of system outlined above, or a mixture of any two, or all three of them. The naive observer has no information about the type of system being observed, and the less-naive observer only has partial information. Any hypothesis formulated by the observer cannot by tested by behavioural methods alone, because for the observer's model to be unique the system would have to be completely observable and completely controllable (see McFarland & Bösser 1993, pp. 145–148).

Therefore, we are inclined to the view that the observer can never really know about the goals of an agent. The agent may know something about its own goals, but not everything. Only the designer can know everything.

5. COGNITION

Since classical antiquity the following tripartite division of the mind is sort of a commonplace among educated western people: The first part is cognition and covers believing, the acquisition and revision of beliefs, learning, knowing and reasoning. Instead of 'cognition' some authors have used words like 'intellect', 'reason' or *nous* (the Greek word for reason). Emotion is the second part and covers feelings, affections and moods. Conation (or volition) covers willing, striving, endeavor, intending, motivation, etc. The often used corresponding adjectival forms of these words are 'cognitive', 'emotional', 'conative' and 'volitional'. The philosophical tradition conceived of cognition as the ability to acquire declarative knowledge, or knowledge that. In other words, to acquire conscious, articulated beliefs, which are true and justified. For two thousands years the paradigm for this type of knowledge was Euclid's *Elements*. This tradition tended to equate all knowledge with knowledge that and all forms of intelligent behavior with behavior guided by knowledge that, i.e., knowledge of facts and knowledge of rules. Gilbert Ryle (1949) attacked this picture as "the intellectualist legend" and pointed out that intelligence is not only manifested in behavior guided by knowledge that, but also in behavior that simply shows know-how. The latter seemed to him to be even more fundamental than the former. His main argument against the intellectualist legend was the following: If behavior is intelligent only when it is guided by reasoning and if reasoning itself can either be done intelligently or not, then reasoning is intelligent when it is guided by reasoning, but then you are trapped into an infinite regress. We often believe and do the right things without having first consulted all relevant reasons and rules. Practice and know-how precede articulated knowledge that. This know-how is fundamental for the rest of our mental lives.

5.1 *Knowing How and Knowing That*

What do we normally mean by know-how? We mean knowing as a matter of knowing how to do something, e.g., how to swim, to play tennis, to climb, to drive, to ride a bicycle, to open a lock, to tie a knot.

Cognate expressions include: skill; craft; practical knowledge; procedural knowledge; expertise; good, reliable performance; behavioral capacity or ability; mastery; habit; automatic behavior; fixed action patterns; reflexes; doing things without reasoning; intelligent behavior without intentional (mental, cognitive, psychological, contentful) states; task-bound knowledge; implicit knowledge; knowledge which does not get transferred to the ability to come to terms with other tasks; knowledge that is not articulated, stated; knowledge for which the having of reasons is irrelevant;

knowledge, that is not language-infected; ability to discriminate and recognise without being able to categorize conceptually: NETTALK, for example, distinguishes between vowels and consonants, but does not posses a task-independent, general or abstract concept of what a vowel or of what a consonant is. This is argued in Clark (1993, pp. 50–53, 70–77).

One can summarize this by saying, that knowledge how is (somehow) in the system in contrast to knowledge that which is available for the system (in the sense of something the system can work on, operate on, make use of in different contexts).

We should mention that there also exist cognitive skills or cases for cognitive know-how: knowing how to find proofs, how to counter objections, how to read. Language use and language understanding are also cognitive skills or cases of knowing how. The ability to produce and to understand potentially infinitely many grammatically well-formed sentences is usually not accompanied by an explicit declarative knowledge of rules.

What do we normally mean by knowledge that? We mean knowing as a matter of having accessible, in different ways usable information about an object, a person, a place, a situation or occasion, an event, a subject matter, etc. Declarative knowledge, understood literally, means: knowledge people declare to possess; what people declare to be the case. In this sense, declarative knowledge can only be had by language-using agents. Evidence that someone has this or that piece of declarative knowledge includes his utterances (the only evidence – according to Davidson (1985) and O'Leary-Hawthorne (1994); but among other evidence – in the opinion of Dennett (1987) and Bennett (1990) and others). Cognate expressions include propositional knowledge, explicit knowledge; having an articulated consciousness, that p; having concepts; having language; having the capacity to exploit given information along many different lines and for many different contexts. Different authors have described this competence in different terms. Fodor speaks of "the isotropy & the Quineian holism" of central cognitive processes (Fodor 1983, pp. 105ff.), arguing that the fixation of particular beliefs is responsive to all beliefs one has. The word 'holism' is often used in this context: Any belief may turn out to be relevant in the course of the epistemic assessment of any other belief. Haugeland speaks of "the semantic intrigue" (Haugeland 1985, pp. 216–217). Stich uses the expression "inferentially promiscuous"; subpersonal, nonconscious knowledge is not inferentially promiscuous (Stich 1978). Pylyshyn uses the expression of "being cognitively penetrable" (Pylyshyn 1984, pp. 130–145). This is connected with the ability to use one and the same information in different tasks and in different contexts; easily accessible and transferable knowledge; knowledge that is susceptible to rational control and critique, to questions of justification, evidence, plausibility, confirmation, consistency, coherence, etc. This is knowledge for the system as contrasted with knowledge in the system.

This characterization of knowledge gives some hints to what a genuine (advanced) thinker might be. A genuine (advanced) thinker can work on its own cognitive and conative states. He can ask himself, whether his own cognitive and conative states live up to the standards he accepts (rationality, coherence, confirmation by evidence, etc.). But one has, on the other side, to accept the limitation on the reach of articulated thought. Every cognitive activity (reading, observation, thought, communication, etc.) requires a lot of processes that will evoke non-intentional, more or less automatic responses. No one who can read is able to state how he recognises words, how he analyses ("parses") sentences, how he comes to understand a text. In other words, there are certain crucial responses required for the cognitive activity of reading that one cannot directly will; one can only wait for them to happen. This is valid for all thought, reasoning, inferences, acts of judgment. Somehow the thinker has to go from If p, then q and p to q (*modus ponens*) without consulting further rules; otherwise he (or she or it) would be victim to infinite regresses, i.e., not really be a thinker. In other words, there have to be mechanisms which ensure that certain reactions are forthcoming, reactions on which the thinker then can operate in a more or less articulated and circumspect manner.

A genuine thinker has the following sort of autonomy: He can work on his own cognitive states. He has reasoning power. He is not stuck with a fixed stock of beliefs and desires. He can acquire new ones, check and modify old ones and give up those which do not conform to the conditions of rationality, plausibility, empirical confirmation, etc. He can search for relevant information: I enter a contract with B only if I find that his record of cooperating with others is a satisfactory one. I can give up opinions if I find them wanting the appropriate confirmation. A genuine thinker is also autonomous in the following sense: he can modify certain aims in the light of other aims he has.

5.2 *Distinguishing Between Procedural, Explicit and Declarative Knowledge.*

Procedural knowledge is knowing how. It is knowledge that is tied to a procedure. It cannot, therefore, be used in another procedure, or be accessed by another process. Dickinson (1985) equates procedural knowledge in rats with habits.

Explicit knowledge involves explicit representations of facts that are (by our definition of explicit representation) accessible to many processes. Explicit knowledge, therefore, involves tokens, which represent 'facts'. Philosophers usually distinguish between knowledge and belief. For example, if someone believes that Paris is the capital of France, then it does not follow that Paris is the capital of France. If someone knows that Paris is the capital of France, then it follows that Paris is the capital of France. Therefore, one cannot know that Lyon is the capital of France, although one may (mistakenly) believe this to be the case. Unfortunately, practitioners of AI and experimental psychology do not make this distinction. They use the term 'knowledge' to cover both knowledge and belief. We feel obliged to succumb to this confusion, in the interests of communication with our colleagues. Therefore, for purposes of this discussion we will use the term knowledge to cover both knowledge and belief.

Declarative knowledge (that people declare) is available only to language-using agents. It is a variety of explicit knowledge that pertains to humans, and possibly to artificial agents in the future. Where Dickinson (1985), and others who have experimented on animals, use the term 'declarative knowledge', we substitute the term explicit knowledge.

Attempts to distinguish empirically amongst these different kinds of knowledge are in their infancy. The question of what kind of evidence would count is controversial (e.g., Bennett 1990; Heyes & Dickinson 1990; O'Leary-Hawthorne 1993). Dickinson and his coworker attempt to experimentally investigate the distinction between procedural and explicit knowledge in rats. Clark and Karmiloff-Smith (1993) address the question of the scope of possible learning in procedural and declarative systems. Their notion of representational redescription gives one a feeling for possible qualitative differences between different types of learning. With language and with cognitive structures learning gets much more powerful than it is on the procedural level.

5.3 *Minimal Cognition*

What is the lower boundary of know-how? "Do you know how to breathe" seems to be a senseless question. But whether it is senseless depends on the context in which it is raised. If we ask a child it will probably demonstrate that it can breathe. Singers, sportsmen have to learn breathing techniques; playing wind instruments requires doing breathing exercises. This indicates that the question "Do you know how to do x?" is appropriate when doing x is something one has to do to accomplish a task.

Do human babies know how to suck? Yes, even though they do it sponta-neously, and it is not something they learn to do. Nevertheless, doing x does not always imply knowing how to do x. It may be that one did x by chance (unlocking the safe by fooling around) or that it was inevitable but not part of a relevant task. Often, one could not have done otherwise as a matter of natural laws. If the robot starts moving around it displaces molecules in the surrounding air; but it does not make sense to say, that the robot knows how to displace molecules in the surrounding air. We say this because displac-ing molecules of air is not a task relevant to the robot. However, displacing pucks is a task relevant to the robot, so we are inclined to say that the robot does know how to displace pucks. Automata can be said to have know-how when they have procedures relevant to some task that the designer intended the robot to accomplish.

What is the lower boundary of explicit knowledge? Explicit knowledge involves explicit representation of facts that are accessible to many processes (see above). At the simplest level it is enough if a single number (the result of some process) is deposited in a store in such a way that it is available to other processes operating within the system. Examples are simple, one-stage planning; symbol manipulation, etc.

The question of how we could know whether an animal or robot is ca-pable of minimal cognition (manipulation of explicit knowledge) is prob-lematical (McFarland 1991; Lanz 2000). It may be that 'only the designer can know', or it may be that empirical investigators can provide criteria by which we could judge.

*Universität Bielefeld, Germany
**Balliol College, Oxford, England

REFERENCES

Bennett, J. (1990). *Linguistic behaviour*. (2nd ed.). Cambridge,UK: Cambridge University Press.

Clark, A. (1993). *Associative engines. Connectionism, concepts, and representational change*. Cambridge, MA: MIT Press/A Bradford Book.

Clark, A., & A. Karmiloff-Smith (1993). The cognizer's innards: A psychological and philosophical perspective on the development of thought. *Mind & Language* **8**, 487–519.

Davidson, D. (1982). Rational animals. *Dialectica* **36**, 318–327; reprinted in E. LePore & B. McLaughlin (eds.) (1985), *Actions and Events* (pp. 473–480). Oxford, UK: Blackwell.

Dennett, D. (1987). *The intentional stance*. Cambridge, MA: MIT Press/A Bradford Book.

Dickinson, A. (1985). Actions and habits: The development of behavioural autonomy. *Philosophical Transactions of the Royal Society of London B 308*, 67–78.

Fodor, J. (1983). *The modularity of mind. An essay on faculty psychology*. Cambridge, MA: MIT Press/A Bradford Book.

Harnard, S. (1990). The symbol grounding Problem. *Physica* **D 42**, 335–346.

Haugeland, J. (1985). *Artificial intelligence. The very idea*. Cambridge, MA: MIT
Press/A Bradford Book.

Heyes, C., & A. Dickinson (1990). The intentionality of animal action. *Mind &
Language* **5**, 87–104.

Holland, O.E., R. Beckers, & J.L. Deneubourg (1994). From local actions to global
tasks: Stigmergy and collective robotics. In R.O. Brooks & P. Maes (eds.), *Artificial Life IV. Proceeding of the Fourth International Workshop on the Synthesis
and Simulation of Living Systems* (pp. 181–189). Cambridge, MA: MIT Press.
Also appeared in H. Ritter, H. Cruse, & J. Dean (eds.), (2000). *Prerational intelligence, adaptive behavior and intelligent systems without symbols and logic,
Vol. 2* (pp. 549–563). Dordrecht, The Netherlands: Kluwer Academic Publishers.

Lanz, P. (2000). Intelligence and prerational intelligence: Introduction to Part II.
In H. Cruse, H. Ritter, & J. Dean (eds.), *Prerational intelligence: Adaptive
behavior and intelligent systems without symbols and logic,* Vol. 1 (pp. 7–18).
Dordrecht, The Netherlands: Kluwer Academic Publishers.

McFarland, D. (1991). Defining motivation and cognition in animals. *International
Studies in the Philosophy of Science* **5**, 153–170.

McFarland, D. (1989). The teleological imperative. In A. Montefiore & D. Noble
(eds.), *Goals, no goals and own goals* (pp. 211–228). London: Unwin Hyman.

McFarland, D., & T. Bösser (1993). *Intelligent behavior in animals and robots*.
Cambridge, MA: MIT Press/Bradford Books.

O'Leary-Hawthorne, J. (1993). Belief and behavior. *Mind & Language* **8**, 461–486.

Pylyshyn, Z. (1984). *Computation and cognition. Toward a foundation for cognitive
science*. Cambridge, MA: MIT Press/A Bradford Book.

Russell, B. (1921). *The analysis of mind*. New York: Macmillan.

Ryle, G. (1949). *The concept of mind*. London: Hutchinson.

Shepard, R.N. (1987). Evolution of a mesh between principles of the mind and
regularities of the world. In J. Dupre (ed.). *The latest on the best: Essays on
evolution and optimality* (pp. 251–275). Cambridge, MA: MIT Press/Bradford
Books.

Stich, P. (1978). Autonomous psychology and the belief-desire thesis. *The Monist*
61, 573–591.

HOLK CRUSE*, JEFFREY DEAN**, and HELGE RITTER*

GENERAL CONCLUSIONS AND OUTLOOK FOR FUTURE WORK

1. INTRODUCTION

This final chapter first reviews different aspects of intelligence that have been covered in the preceding chapters. It revisits the distinction between rational and prerational intelligence that has been proposed here. It considers significant characteristics of systems exhibiting prerational intelligence and identifies several areas where further research is needed. It also reflects on some of the similarities and differences in the perspectives from the different disciplines. The chapter concludes with the major, to many readers perhaps the most intriguing, open question – the place of our subjective experience in intelligence.

2. RATIONAL AND PRERATIONAL INTELLIGENCE

In the course of the book we have seen that the word "intelligence" can be used in several different ways. For an external observer with no access to the internal workings of the system, any judgement about intelligence can be made only on the basis of the observable behavior. As Lanz and Mc-Farland (see *Philosophical Perspectives*) suggest, such a judgement must remain at the level of a hypothesis except in the very restricted circumstances of a fully controllable and fully observable agent. Most chapters (e.g., Dean et al. (*Biological Perspectives*); Lanz & McFarland (*Philosophical Perspectives*); Braun (*Computer Science Perspectives*) have considered at least some attributes that influence this judgement. One common criterion is that intelligent behavior have utility for the agent. For animals, where maximizing reproductive fitness provides a clearly defined goal or evaluation function, intelligent behavior is closely related to adaptive behavior – that which improves fitness. This criterion, modified to emphasize survival and maintenance of function rather than reproductive fitness, has been extended to the evaluation of autonomous robots. However, this focus on utility clearly ignores several important attributes of human intelligence and the concept of intelligence as it is used in everyday life. Furthermore, it does not help to differentiate intelligent behavior from simple forms of adaptation or to explain several characteristic features of intelligence – features like ro-

J. Dean et al. (eds.), Prerational Intelligence: Interdisciplinary Perspectives on the Behavior of Natural and Artificial Systems, 265–283.
© 2000 *Kluwer Academic Publishers.*

bustness, flexibility, and applicability to complex and novel situations – that are mentioned in many of the perspectives.

Attempts to include one or more of these characteristic features have led to many definitions of intelligence and to many attempts to categorize different types of intelligence. None has been universally accepted, causing some psychologists to conclude simply that "intelligence is what is measured by an IQ test" (e.g., see *Philosophical Perspectives*). One fundamental problem with most attempts at a unitary definition of intelligence may be the neglect of the constraints under which biological intelligence has evolved: intelligent behavior must advance the fitness of an animal in its own particular habitat and niche. Thus, an important qualification on utility as a criterion of intelligence is that the utility of a behavior depends upon both the animal and the context. Intelligence is manifest not in any particular behavior alone but rather in a behavior performed in a specific context. Given the diverse evolutionary histories and current niches of existent species, one should not assume that all the demands of animals' complex interactions with their natural environment are met best by a single universal intelligence.

Most of the authors represented here accept that intelligence is multifacetted and that rational intelligence is best regarded as simply one mechanism for achieving intelligent behavior. This, of course, raises the question of when rational intelligence is the better method – a question considered in several chapters (e.g., *Computer Science Perspectives* and *Psychological Perspectives*).

Learning, an important component of intelligence, provides a model in this respect: learning was previously thought to be a single, universal capability that could be applied to a greater or lesser extent in different situations. It is now clear that several different kinds of learning can be dissociated according to their contexts and attributes as well as the neural substrate upon which they depend. It is also clear that several different forms of neural plasticity contribute to the observable behavioral changes defined as learning (see *Biological Perspectives*). In a similar way, intelligence is probably best regarded as possibly distinct capabilities linked to different contexts and not necessarily dependent on a single intelligent process.

The growing appreciation of different forms of intelligence provided the primary motivation for organizing the research group that led to the present book. The particular motivation was the realization that the facility most readily associated with intelligence, that of abstract logic and symbolic reasoning, was poorly suited to some apparently simple, everyday tasks. Solving mathematical problems, playing chess or similar high-level, abstract tasks typically come first to mind as examples of highly intelligent human behavior. However, our daily life, and even more so that of any animal, is filled with perceptual and motor tasks that are quite different and turn out to be difficult to solve for artificial systems relying on logical reasoning. In fact, as the analysis of nervous systems and the success of animats and artificial neural network research show, these tasks do not require logical reasoning and they may be better performed by systems that do not utilize it. Because these so-called low-level systems, which we share with animals, provide the inputs and outputs to the logical and reasoning facility of humans, we presume that the function of the former enables and strongly constrains the expression of the latter. As a consequence, we further argue that the phenomenon of intelligence can only be fully understood in conjunction with an understanding of these comparatively simple processes that we subsume under the term prerational intelligence. The hope for successfully achieving this understanding lies in the rapidly multiplying examples of emergent properties – properties and behaviors that arise from non-linear interactions among the components of a system. Although emergent, qualitatively new properties can be found even in simple systems, their richness generally increases with the complexity of the system, leading to a crude continuum of increasing complexity of structure and intelligence of behavior. In the *Psychological Perspectives*, Bridgeman et al. discuss a continuum of processes within an individual; In the *Biological Perspectives*, Dean et al. sketch a continuum in terms of autonomous agents of increasing complexity. Within the latter continuum, stretching from the clearly reactive to the clearly intelligent, different observers presumably will always favor different thresholds for accepting a system as "intelligent", but agreement on a particular threshold is not necessary for progress towards understanding the general phenomenon.

Before further consideration of this continuum, the two qualitatively different uses of intelligence and their relation to the observer must be reviewed: in the terminology of the Research Group, the attribution of intelligence to an agent may reflect either an adverbial use or a nominal use. A system might behave intelligently in the judgement of an observer – this is the primary adverbial sense – because it possesses appropriate structure and procedures that have been "intelligently designed" either by its builder or by evolution – this is by extension another adverbial use – or because the system itself possesses the capacity of intelligence, meaning some measure of logical thought and symbolic reasoning – this is the nominal use meaning that the internal procedures themselves explicitly incorporate features colloquially associated with intelligence.

In biological contexts, the adverbial use would be identified with "species intelligence" in a narrow sense – the set of innate responses and behaviors acquired by a species through evolution, primarily specified by the genes and therefore "hard-wired" more or less identically in the nervous system of each individual. More generally, the term "competence" has sometimes been used to designate such abilities tied closely to a particular task. In contrast, nominal intelligence might be loosely identified with individual intelligence going beyond the natural variation in the genetic specification, that is, intelligence resulting from the interaction of the individual with its environment and linked to individually acquired representations or procedures. Another way to look at this distinction would be to say that adverbial intelligence shows a solution to a problem that has already been solved earlier, whereas nominal intelligence is necessary to find a solution to a new problem.

Additional distinctions and terminology follow from this observation. Lanz and McFarland's discussion of implicit and explicit representations and their corresponding limitations are one example (see *Philosophical Perspectives*). More generally, we say that systems with nominal intelligence possess intelligence of content or rational intelligence, whereas those with adverbial intelligence possess intelligence of design or process, regardless of whether the designer is a human – for machines and other artifacts – or biological evolution. Systems that possess intelligence of design but not necessarily intelligence of content exhibit that which we call prerational intelligence. Hence, we identify intelligence primarily with aspects of behavior and link the adjectives rational and prerational with aspects of mechanism.

From this description it follows that intelligence evident in behavior implies at least some measure of adverbial intelligence, that is, intelligence of design or process; it may, but does not necessarily, reflect nominal intelligence, intelligence of content. Conversely, poorly realized intelligence of content may lead to behavior not characterized by intelligence and, more rarely, one might judge a system to be intelligently designed to do what it does, even if the behavior itself would not be considered intelligent in an adverbial sense.

Beyond these observations, however, the natural relationship between mechanisms supporting rational and prerational intelligence is a major open question. According to the traditional view, rational and prerational would represent two exclusive domains. In the *Computer Science Perspectives*, Braun suggests that mechanisms of prerational intelligence correspond to special-purpose machines that work fast but rely on dedicated and therefore in some sense costly hardware whereas mechanisms of rational intelligence correspond to less costly general-purpose machines that necessarily work more slowly. This corresponds to the implicit/explicit dichotomy discussed by Lanz and McFarland (see *Philosophical Perspectives*), the notion that explicit representations or, by analogy, mechanisms of rational intelligence are open-ended in the sense that they can be applied to tasks other than those for which they originated. Braun's suggestion raises the possibility that prerational and rational mechanisms might be distinguishable on the basis of observed performance if temporal and spatial constraints can be taken into account. This would provide a way around the strict requirements on observability and controllability, but only if the two categories of mechanisms are sufficiently distinct in performance. This is uncertain, even in technical spheres where clear examples of one or the other can be constructed. Instead, the dichotomy may itself turn out to be more of a continuum. From the perspective of biology, where the presumption is stronger that the underlying mechanisms are shared, we tend rather to the view that rational intelligence represents the expression of a subset of prerational mechanisms. For biologists, lacking the possibility of creating their material, the crux of the problem is to discover how neural structures and dynamics are shaped to create specialist machines in some cases and generalist machines in others, a point we will return to below.

Thus, knowledge of the behavior alone does not permit a distinction between adverbial and nominal intelligence. The particular attribution of intelligence hinges on the observer's detailed knowledge of the workings of the system in question, as Lanz and McFarland (see *Philosophical Perspectives)* consider at length. One must distinguish between the behavior and the mechanism that produces it and also distinguish between knowledge of the external and the internal workings of the system. Only an observer with complete access to the inner workings of the system can make the required judgement. (This access should be distinguished from the internal aspect, the limited access to our personal mental processes to be discussed below.)

This requirement for knowledge of the inner workings creates a predicament. On the one hand, we can clearly distinguish according to mechanism only those artifacts that humans design and construct, artifacts that do not yet exhibit convincing measures of what is commonly felt to represent intelligence. (Even here our confidence in our full understanding of man-made systems generally decreases as system complexity increases and further decreases for artifacts generated in part via self-organizing processes.) On the other hand, we do not yet have nearly enough physiological information about complex animals to understand how their intelligent behavior comes about. Despite these inadequacies, research in the fields surveyed here suffices to provide numerous examples of systems that behave intelligently in the more or less convincing absence of nominal intelligence. Moreover, good performance in some tasks can be achieved with either prerational or rational mechanisms. In other cases, such as motor learning, rational mechanisms typically play a role in tuning prerational mechanisms: improving athletic performance under instruction may involve a period of heightened conscious or voluntary control leading finally to a higher level of automatic performance. These relationships support the notion of a close bond between rational and prerational intelligence, rather than a clear qualitative disjunction.

We are confident that the intelligence of content that we believe to exist in humans must rest on an intelligence of design arising through evolution. This is the basis for our belief that mechanisms supporting rational intelligence will turn out to be closely related to those involved in prerational intelligence.

As a corollary, we expect that the transition to intelligence of content will not be a qualitative disjunction but rather a matter of quantity and complexity. Therefore, we will begin with a review of characteristics and open questions associated with prerational intelligence and then consider possible extensions to rational intelligence. Realizing that particular implementations for achieving intelligent behavior differ, we will next revisit those features that might characterize general classes of intelligent behavior and the mechanisms that support them.

3. CHARACTERISTICS OF SYSTEMS SHOWING INTELLIGENT BEHAVIOR

Perhaps the single characteristic that best defines intelligence on the behavioral level is the ability to behave appropriately in new situations, that is, to solve problems that are novel in some sense. (Obviously, for novel problems to be tractable, they must be part of a more or less familiar class of problems or share features with previously encountered problems.) We adopt this focus here. The parallel goals of the inquiry, then, are to understand how animals are able to do this successfully and to construct machines with similar capabilities.

Of course, improvements in adaptation and reproductive success are the ultimate reasons for the acquisition of these capabilities by a species in the course of evolution. Intelligence and intelligent behavior must improve the success of the constituent individuals in acquiring resources, surviving and reproducing. Responding appropriately to novel situations is one component of this ability. Examples of this species intelligence, like many other aspects of biological adaptation, often can be labeled usefully as representing either a "specialist" or a "generalist", although these distinct cases are more likely to represent ends of a continuum than clearly separate categories. Properly used, the terms reflect the behavior of the individual or species in relation to the context and the corresponding set of potential behaviors; they are not inherent properties of the individual itself.

On the one hand, specialists rely on highly evolved adaptations to a narrow niche, where the niche is defined as the multidimensional space of physical and biological parameters within which the animal can survive. An example might be exquisite and efficient procedures for utilizing a specific food source. Parasites dependent on finding a single host species would be an example. A more complex behavioral example would be carnivorous female lightning bugs of one species employing one code to attract conspecific males for mating and the code of another species in order to feed on the duped males. The essence of a specialist is the precise adaptation to a particular, narrow task and the efficient performance based on fine-tuning sensors, central processing and motor output for this particular constellation. If the appropriate constellation occurs frequently enough to ensure the reproductive success of the specialist, its poor or inappropriate behavior in other constellations becomes irrelevant, so specialists can give up the adaptability to a wide range of novel situations. Specialist strategies can be enormously successful in evolution, but once we understand the basis of such a specialization, we are apt to consider it to show intelligence of design, but to lack the flexibility and variability characteristic of intelligence. Many machines, especially the special-purpose machines referred to above in connection with prerational mechanisms, can be considered successful specialists.

On the other hand, generalists are characterized by extremely large niches. They perform well in a wide range of physical and biological situations. One strategy enabling a species to be a generalist is to maintain great behavioral flexibility in the individual. This strategy represents species intelligence in the broad sense that the genetic specifications merely provide general procedures and problem-solving routines enabling the individual to response adaptively to many different, novel situations. Learning, developmental plasticity, and the problem-solving abilities of higher primates and of course humans represent generalist strategies. It is this ability to respond adaptively to environmental change by adaptively switching strategies either in response to failure or as a result of internal processing that is at the heart of the mystery of intelligence. Machines with this kind of flexibility in behavior are not yet available.

In species adopting this generalist approach to adaptation, increasing flexibility of behavior is associated with increasing complexity of the neural control systems.

Recent research in various fields is providing better proximate explanations for how intelligent behavior is achieved in the individual. Analytic research in the realms of biology and psychology, discussed here in the *Bio-*

logical Perspectives and *Psychological Perspectives* (but see additional examples collected in Cruse et al. (2000) and Ritter et al. (2000)), provides numerous examples of intelligent behavior that rely on processes that are not rational or symbolic in any conventional sense, that is, do not possess intelligence of content. To recall just a few examples, we could mention the collective intelligence of ants, the subconscious processing of visual information to make judgements of recognition or to control movement, the coordination of multiple rhythm generators, and the allotment of time and resources by animals.

Synthetic research in fields of computer science, psychology, and artificial intelligence – excluding good old-fashioned AI – also makes clear that intelligent behavior can be achieved without explicitly using symbols and logical thought (e.g., Cruse et al. 2000; Ritter et al. 2000). Artificial neural networks can produce good approximations to arbitrarily complex transformations; they generalize appropriately; they are robust with respect to incomplete information or component failure. Still other artificial neural networks indicate how networks of competing and cooperating modules can be used to structure behavior over time into sequences that are useful relative to purposes defined explicitly or implicitly by a designer. Examples of autonomous robots, primarily drawn from basic research projects, provide some initial if still rather elementary examples of how machines built along such lines can exhibit unexpectedly complex behavior, both individually and collectively. The designers of these man-made artifacts can verify that intelligence of content is absent, at least as long as self-organizing processes play a relatively minor part.

Designers' experience with artificial networks of different architectures indicates what features might correlate with increasingly intelligent performance. To date, systematic knowledge of the relationship between structure and functional capabilities is still rudimentary, but distributed representations and parallel processing obviously characterize these successful artificial networks.

Ever-increasing knowledge of biological nervous systems exposes analogous properties: distributed representation and parallel processing are typical of biological control systems responsible for all but simple stereotyped behaviors. As outlined in the *Biological Perspectives*, adding recurrent connections to feedforward networks greatly increases their capabilities. Architectures based on cooperation and competition are found at both low-levels and high-levels of processing. For example, the selection of the behavior or behaviors to perform at a given time is accomplished by a system of cooperating and competing modules, where a module corresponds to a command neuron or sets of command elements (neurons) making up a command system that switches a particular behavior on or off. (The artificial networks pioneered by P. Maes (1991) implement the same architecture and replicate many features described in classical ethology; Dörner and Gerdes (2000) present further developments.) Many studies in behavioral ecology indicate that animals allot time and energy as if they were performing an analytic optimization, but neurophysiological results in analogous systems show no evidence that an explicit optimization is performed. In fact, examples in several spheres indicate that animals rely on approximate algorithms that suffice for adaptive behavior but are simpler to implement in the neural hardware than exact algorithms.

Thus, several chapters provide a basis for optimism that the notion that prerational mechanisms provide a good measure of intelligent behavior does apply to many capabilities of animals and that it will prove useful for obtaining intelligent behavior in machines. Various learning algorithms for individuals and genetic algorithms for populations have achieved astounding success in a relatively short time. The necessity for intelligence of content to achieve high-level intelligence has not been demonstrated, despite our introspective intuition to the contrary. Humans taking part in an intellectual discussion on a mathematical problem, to take an example of an application of rational intelligence, are not able to explain how they can recognize or formulate the words and sentences used. Thus, rational mechanisms appear always to be based on prerational processes; to use the words of Lanz and McFarland (see *Philosophical Perspectives*), know-how precedes articulated knowledge. This conclusion is still more convincing when we consider motor intelligence (see also the sections of the *Psychological Perspectives* by Velichkovsky and Paillard), which encompasses a large category of problems that appear to be easy but turn out to be resistant to rational solutions relying on symbolic processing. Examples are riding a bike or skiing or many problems that deal with the degrees of freedom problem.

4. CHALLENGES FOR THE FUTURE: CONSTRUCTING INTELLIGENT SYSTEMS

The *Perspectives of Engineering and Robotics*, however, make plain that, despite the success of initial attempts to imitate biological control systems, many of these innovations in basic research do not yet meet the demands of the market place with the possible exception of specialized niches like toys (Grand 1998). With some exceptions, traditional control techniques are still superior in various respects. At least two reasons might be advanced for this failing. One, the problem of controllability, is a problem inherent in many applied tasks. The very features characteristic of intelligence – that it can deal with new situations and find new solutions to old problems, that it is complex and shows emergent features – make it unsuitable for applications where the behavior of the system must be predictable at least in so far as a judgement about safety can be made. What the market requires is limited autonomy, in particular limitations on what McFarland and Bösser (1993) refer to as motivational autonomy.

The second problem is that artificial systems are still far from the level of complexity existing in biological nervous systems. According to some accounts, numbers of elements are achieving parity, but the heterogeneity and modularity within biological control networks is far greater. Presumably, these features are necessary prerequisites for the emergence of a truly generalist intelligence that can deal with new situations and adapt in a useful way. The comparatively primitive behavior exhibited to date in fully autonomous robots supports this view. Low-level behaviors like wall following, obstacle avoidance, and pursuit have often been achieved and they can "emerge" from still more basic behaviors. However, even the task of navigating under optical control in a novel environment and building up some sort of representation of the environment, a basic ability we take for granted in ourselves and other animals, is still incompletely solved.

Nevertheless, we are confident both these problems can be mastered with increasing understanding of how properties emerge in complex systems and with improvements in technology and biological understanding. Clearly, modularity is one of the major open questions concerning prerational intelligence. Biological networks show a much higher degree of modularity than artificial systems do, and they show more characteristics associated with intelligence, either in quantity or quality. Are these facts related? Modularity is also prevalent in many areas of design and engineering, but in these fields it seems to serve the purpose of making a system predictable by breaking it into analyzable pieces that communicate with each other in restricted, well-defined ways. One goal of modularity in engineering is to avoid surprises, or emergent phenomena. Biological modularity, because the purpose is to increase the fitness of the whole system, may have followed a different strategy, one that opportunistically uses emergent phenomena as long as they increase fitness. Furthermore, modularity in each animal is a result of a long evolutionary history, and it is very much unclear how much of the structure is determined by functional necessity and how much simply by chance and by constraints of development and evolution. If this is so, how can we learn to properly use modularity and set boundaries on emergent phenomena? Each animal, of course, has much of its modularity already fixed in its genes and the development they specify; engineers and designers do not have this head start.

Several general problems require solution in a modular system. One is simply how to structure transmission of information among modules. Another is how to enable just one or an appropriate combination of all candidate modules to control behavior; this specific problem occurs in behavior-based architectures where different modules represent different behaviors, but it occurs any time modularity is structured horizontally with multiple units present on the same functional level. Action selection is one context; memory retrieval is another example. A third problem is related to credit assignment questions in learning and memory storage where the problem is to determine whether a module should be changed and, if so, which one. A fourth problem concerns cooperation of modules. How modules can be combined dynamically to form different structures on a higher level? How can modular structures be used to form and manipulate the functional equivalent of a world model, that is, a unified representation combining knowledge of facts and their relationships? Finally, there is the question of creating novel structure and functionality in modular structures, that is, building new combinations of previously constructed elements?

4.1   *Challenges for the Future: A Unified, Interdisciplinary Theory of Intelligence and Intelligent Systems*

Beyond the practical challenge described in the previous section, the reconciliation of the diverging viewpoints that are evident at several points in the different chapters remains a major challenge. Some of these diverging views reflect deeply ingrained traditions within the different fields. It is perhaps not surprising that the deepest division is between psychology, with its primary focus on human behavior and high-level, cognitive functions, and most of the other fields, with their low-level, mechanistic approaches. As the psychologists themselves explain, the approaches subsumed under prerational intelligence encroach upon traditional spheres of psychological inquiry (see *Psychological Perspectives*). At its core, psychology has more in common with the top-down analyses of linguistics and classical AI than with the low-level focus and bottom-up approaches explored under the label of prerational intelligence. Hence, it is not surprising that some psychologists are uncomfortable with the notion of prerational intelligence.

To some extent, the discrepancies are a question of how to deal with incomplete understanding. For example, hypothetical representations – mentalistic constructs variously called plans or schemata – play a central part in psychological models. Plans in the everyday sense, of course, can be readily expressed in words, so this explanatory metaphor is intimately connected to our rational intelligence. The concept of a schema does not carry such a strong rational implication, if only because the word is less common and possible underlying mechanisms are rarely mentioned. Hence, schemata are potentially open to prerational interpretations, but to date they still seem to be linked to logical constructs. In summary, both metaphors serve a useful descriptive purpose given the present level of knowledge, but plans and schemata ultimately need to be explained in terms of neural structure and function, the level addressed by biology and computational neuroscience. As increasing knowledge of neural function is acquired, the perceived gap between the levels addressed by psychological theories and by biological theories of brain function should naturally disappear. Of course, acquiring deep understanding of how such high-level attributes emerge from the low-level dynamics of neural networks represents a formidable challenge.

Similarly, psychologists are more likely to want to identify any prerational-rational distinction with one or another of the dichotomies that have a long history in the field (see *Psychological Perspectives*). These include unconscious versus conscious processing, automatic versus voluntary, reactive versus predictive control, or action versus thought. However, none of these dichotomies precisely reflect the distinction being developed here. In particular, as discussed below, we do not want to identify consciousness as a hallmark of rational mechanisms and consciously impenetrable as a necessary feature of prerational mechanisms. Among other problems, this identification would raise the question of whether the unattended control processes preceding a period of conscious, rational effort differ qualitatively from the automatic control finally achieved, or as one contributor suggests (see *Psychological Perspectives*), is there a post-rational intelligence achievable through rational intelligence?

Despite these differences the *Psychological Perspectives* provide ample evidence that the notion of prerational intelligence can be extended into psychological modeling. First, the importance of unconscious processing is increasingly recognized. Second, Cellérier underscores the important role of interactions with the surroundings that is a key aspect of Piaget's theory, and like the biologists, he emphasizes the combination of genetic and epigenetic adaptation with their different time scales in molding present-day living organisms. He argues against a simplistic identification of prerational and rational with stages in the Piagetian theory, an argument we would share based on the holistic view that intelligence of process ultimately supports intelligence of content. Third, the special tie between language and rational intelligence, which many authors acknowledge, would be weakened if Bridgeman's suggested close link between language and general planning capabilities is accepted. If planning turns out to be derived from prerational mechanisms, rational intelligence would then be just a subset of prerational intelligence that happens to be hypertrophied in humans. Pursuing this idea further, we could point to Bridgeman's discussion of rational thought and recurrent connections for internal language feedback. Recurrent connections, of course, are a major element in many real and synthetic prerational control networks. Thus, combining Bridgeman's idea with ideas from the *Biological Perspectives*, one could speculate that weak recurrent connections originally involved in controlling accuracy of performance in the manner of reafference could take on the larger, more independent role of reactivating circuits for planning. Fourth, several contributors to the *Psychological Perspectives* also seem to support the idea of rational intelligence being a kind of general-purpose mechanism augmenting special-purpose mechanisms that are used in key areas.

A second major gap needing to be closed is that between formal theoretical treatment and biological reality. On the one hand, the *Computer Science Perspectives* and *Mathematical Perspectives* both discuss a variety of formal techniques for measuring relevant quantities like complexity, computability, termination, mutual similarity, information content, and theories of dynamical systems. The chapter by Braun elegantly demonstrates the advantages of a broad interdisciplinary comparative approach. In particular, it shows how computer science addresses problems at the heart of the discussion of intelligence: those related to problem-solving algorithms and their efficiency. It demonstrates the advantages of treating the brain as just another computing machine for comparative purposes, as in the discussion of tradeoffs between memory and processing units. At the same time, the simplifying assumptions, particularly those related to homogeneity that are currently necessary to formalize notions of complexity, are quite unsuited to biological brains, as explain the *Biological Perspectives*. Thus, this refreshingly comparative approach does lead to the intuition that rational intelligence has a special role besides the role served by special-purpose hardware solutions based either on memory or on procedures. Nevertheless, there is still much work to be done before the formal treatments of complexity and information theory can realize their full potential application to real nervous systems.

5. POSSIBLE LIMITS TO PRERATIONAL INTELLIGENCE

Thus, we see that there is still much to learn about prerational mechanisms and their properties. They can accomplish many tasks but do they have limits? In particular, what can we say about the second kind of intelligence, the intelligence of content that we believe ourselves to possess as human beings? Does it lie within the realm of prerational mechanisms? The *Biological Perspectives* strongly suggest an affirmative answer. The basic elements of the nervous system certainly seem to use prerational procedures, so the question is how rationality can arise on this basis? Is there a sharp boundary, some principle difference in the processes and structures necessary to support rational intelligence? Have biological systems implemented the single, logically necessary structure for such behavior or do they largely reflect evolutionary accidents, which would imply that quite different architectures are also possible?

Introspection provides one version of what happens when we believe we apply rational procedures. First of all, rationality is always "bounded". Our rational processes have neither unlimited precision nor unlimited time. Instead, modern psychological research suggests that most of our mental processes can be better described as searching for the best fit (or correlation) rather than applying strictly rational rules. Good examples are the "cognitive illusions" described by Tversky and Kahneman (1986), Gigerenzer and Goldstein (1996), and Gigerenzer (2000). However, we also possess mechanisms for checking consistency, mechanisms that again check some kind of correlation but in this case a correlation with earlier knowledge. If a consistency-checking module detects a contradiction, it elicits a new search for a solution. This formulation raises obvious analogies with technological algorithms that have been discussed in previous chapters. The problem is to uncover how these properties are implemented or approximated in biological systems or, in other words, to trace them back to prerational mechanisms. Cultural evolution, of course, has added a deep layer of symbol-mediated knowledge and tools that we have learned to apply in specific, difficult situations. Such tools might be logical rules, or mathematical procedures, such as for example, formal theories for estimating a probability. In some cases, formal theory and natural intuition seem at odds. (A simple example is the common expectation concerning the outcome of, say, the tenth coin toss, if the previous nine tosses have produced heads.) Rather than focusing on the inadequacy of the apparent misunderstanding of the situation – an evaluation involving specific ideal assumptions, perhaps more attention should be directed to considering how such formally codified abilities are used in the real world. In other words, focusing on the interaction between the agent and its natural environment may well produce a deeper understanding of natural intelligence than comparing the behavior to the abstract standard of formal logic.

The treatment of symbols in a system consisting of neurons or artificial neuronal networks is obviously a key question in this respect. Intelligence of content seems to necessitate the use of symbols as part of explicit representations. As several chapters have emphasized, a representation is not explicit per se, but only with respect to how the information is used or can be used.

According to Ritter and Bauer (*Mathematical Perspectives*), explicit means easy to obtain, whereas Lanz and McFarland (*Philosophical Perspectives*) emphasize the open-ended applicability. According to Deacon (1997), it is not our high intelligence that allows us to use symbols, rather the capability of our brain to use symbols, i.e., to summarize a set of properties by a single token that can be used in different contexts, increases our intelligence. But what is the origin of this power of the human nervous system to create, modify and process symbols, or more precisely, to behave in a way that lends itself to being so described? Surely, increasing understanding of the dynamics of neural interactions in complex networks will eventually link symbols and their use with specific activity patterns, such as attractors and their transitions, in the language of dynamics. With respect to machines, the questions to be answered are can symbols used by machines, say in AI computer programs, be attributed meaning in the same sense as those used by humans or, conversely, can biological principles be implemented in machines?

The present situation reminds us of the discussion during the first decades of this century regarding the phenomenon of life. The difference between dead and living matter was believed to be a difference in principle. There was no way to imagine that dead matter could be infused with life. Special properties, such as the *vis vitalis* or entelechy, were hypothesized in order to explain this qualitative difference. Similar problems occurred in chemistry with fire and phlogiston. Nowadays we have no intellectual problems with these qualitative and apparently radical transitions. We need neither phlogiston nor a *vis vitalis*. We understand how the properties of life arise from mechanisms and components that by themselves are not living; yet the combination is self-sustaining and self-reproducing. We believe this although we are not (yet) able to construct a living system from its raw materials.

Our proposal is that we are confronted with a very similar situation and will have a similar experience in studying the phenomenon of intelligence. At first sight, particularly with respect to more complex forms of human intelligence, we have the impression that this is a very special phenomenon that we experience, and we cannot understand how it occurs. As a result, until recently it was considered obvious that artificial systems cannot be truly intelligent, because they lack insight and cognition and these are properties that one either has or does not have, that is, they are not a matter of degree. Many scientists and philosophers (e.g., Searle 1990) still hold this view.

In contrast, the view taken here is that the study of prerational intelligence will provide a basis for understanding rational intelligence and that the differences in the mechanisms of the two will be a matter of degree rather than one of quality. Any dividing line will be somewhat arbitrary, similar to the way one may argue whether a virus or a prion is a living system depending on different definitions of life. In both analytic and synthetic research, we have seen that examples of processes that can be described as cognitive – in that local processes are used to create hypotheses about global relationships that in turn influence the workings of the local processes – occur at levels besides conscious, logical thought. We should not be misled by our introspective experience into thinking that this experience also reveals the processes by which rational thought comes about. Ratio is definitely not a central instance through which all activity must pass before affecting actions. In other words, it is not a necessary part of a causal chain leading from inputs to outputs – the view implicitly or explicitly assumed by some psychologists (see *Psychological Perspectives*) and exemplified by variations of the sense-plan-act cycle hypothesis.

6. INTROSPECTIVE EXPERIENCE AND PRERATIONAL INTELLIGENCE

For many people, the most intriguing if, from a practical standpoint, not the most important open question concerns the phenomenon of the internal aspect, that is, our private, subjective experience of own perceptual, rational, emotional, and motor processes or in other words what it feels like. Our focus on prerational intelligence emphasizes observable behavior and mechanisms involving emergent properties in systems of simple elements. To date, most examples considered have been low-level processes far removed from anything like subjective experience.

One possibility is that the internal aspect is also an emergent phenomenon but one with no causal role. In other words, the experience of an internal perspective might be an epiphenomenon in the sense that we can observe it, but it does not play a role in the causal chain leading from inputs to outputs. In this view, that which is experienced reflects the processes determining behavior but has no means of influencing them. The internal aspect would be considered a phenomenon belonging to a different level of description.

Bridgeman (see *Psychological Perspectives*) places conscious experience at the site where stored memories are reactivated and used in executing plans. Like many, we presume that the processes that provide the physiological basis of our introspective experience are essential parts of the causal chain leading from sensory input to behavioral response or at least that many if not all components of these processes contribute to this causal chain. This contribution provides a functional rationale for the processes but not for the subjective experience per se. Thus, major questions are still open. First, does the subjective experience or internal perspective have a function? If so, is the function related to intelligence? In particular, does it enable qualitatively or quantitatively new levels of intelligent behavior? Is it necessary for certain levels of intelligent behavior? Second, is it a phenomenon that inevitably arises in certain kinds of systems and, if so, what kinds of systems or conditions are needed for it to occur? As a corollary, is an artificial system conceivable that would experience an internal perspective?

Dennett (1991) has argued that conscious experience has developed as a kind of internalized conversation with oneself, a way of using the advantages of symbolic thought, which would have originated in order to communicate with others, for enhancing one's own mental capabilities. This is a classical preadaptation in a biological sense. The improved mental capabilities would provide a functional role for self-awareness, which would explain its selective advantage. However, like other examples of biological evolution, its present nature could be related more to accidents of biological evolution than to functional necessity. According to the latter interpretation, the path via communication with other individuals might be merely one evolutionary path to particular functions that could be achieved in other ways; if it is correct, the end-product, which includes subjective experience, could be replaced by alternative mechanisms not including subjective experience.

In Dennett's view, what an individual reports is only one, possibly edited, version of events among multiple versions continuously developing in parallel within the brain. Coherency in the different streams is to some extent imposed by external referents and to some extent by interconnections within the brain; coherency in the expressed behavior, of course, is also constrained by the competition for shared motor systems. Sociobiologists would further suspect that evolutionary selection has influenced self-awareness and the type of activity that survives any editing process in order to enhance the individual's fitness in interactions with others. Nevertheless, many would argue that any such mechanism could in theory be replaced by equivalent mechanisms, in other words a zombie, performing the same behavior with equal conviction but without subjective experience. From a biological perspective, one could only respond that in many spheres of behavior evolution attaches importance to the veracity of the behavior's reflection of the performer's motivation. The internal aspect would be subjected to the same evolutionary checks and balances; through grounding in basic physiological needs it may provide an important calibration for evaluating one's own actions and those of others.

Dennett's view does contain elements common to the view presented here: in the language of the Research Group, the brain is a dynamical system subject to many external and internal influences and to the dynamics inherent in its own pattern of connectivity and reactivity. Subjective experience might be a necessary concomitant of the dynamics of such a complex system incorporating many different scales and dimensions. Subjective experience might represent the binding of stimuli and internal state into one unit, creating a common currency to enable the global optimization that biologists denote as fitness. Just how such a sense of self might contribute to heightening intelligence is a subject for further consideration, as is the question of whether it would necessarily arise in artificial systems without full autonomy and the necessity for self-reproduction.

In summary, with the possible exception of the last question, the reviews of the various fields lead us to the conclusion that no fundamental problems obstruct the path to understanding the phenomenon of intelligence. We have the impression that, in principle, we can understand how intelligence might arise in increasingly complicated, but in principle understandable, circuits. Similarly, we have glimpses of how logical reasoning might occur in complex neural networks. This does not mean that at this moment we are capable of fully understanding animal intelligence or building an intelligent artificial system. Many questions remain open. These include both matters of detail as well as major needs for tools to handle the dynamics and emergent phenomena of complex systems. Nevertheless, a *vis intelligens*, for which the term cognition often seems to be (ab)used, is surely unnecessary. Minsky (1987, cit. in Lanz 2000) believes that intelligence is like the former unexplored regions of Africa: it disappears as soon as we discover it. This change in perception has accompanied many past achievements in artificial intelligence. Nevertheless, we are confident that the complexity of the human brain will suffice for us to retain our sense of wonder even as we dissect and understand the prerational mechanisms in the dynamics and self-organization of the brain and body that support intelligent behavior.

*University of Bielefeld, Bielefeld, Germany
**BGES, Cleveland State University, Cleveland, Ohio, USA

REFERENCES

Cruse, H., H. Ritter, & J. Dean (2000). *Prerational intelligence: Adaptive behavior and intelligent systems without symbols and logic*, Vol. 1. Dordrecht, The Netherlands: Kluwer Academic Publishers.

Deacon, T. (1997). *The symbolic species. The co-evolution of language and the human brain.* London: Allen Lane The Penguin Press.

Dennett, D.C. (1991). *Consciousness explained.* Boston: Little, Brown and Co.

Dörner, D., & J. Gerdes (2000). Psi: Eine Theorie der Handlungsregulation. Bern, Switzerland: Verlag Huber.

Gigerenzer, G. (2000). *Adaptive thinking: Rationality in the real world.* New York: Oxford University Press.

Gigerenzer, G., & D.G. Goldstein (1996). Reasoning the fast and frugal way: Models of bounded rationality. *Psychological Review* **103**, 650–669.

Grand, S. (1998). *Animats with attitude: Creating a personality from simple components.* Keynote lecture given at the 5th International Conference of the Society of Adaptive Behavior. Zurich.

Lanz, P. (2000). The concept of intelligence in psychology and philosophy. In H. Cruse, H. Ritter, & J. Dean (eds.), *Prerational intelligence: Adaptive behavior and intelligent systems without symbols and logic,* Vol. 1 (pp. 19–20). Dordrecht, The Netherlands: Kluwer Academic Publishers.

McFarland, D., & T. Bösser (1993). *Intelligent behavior in animals and robots.* Cambridge, MA: MIT Press.

Maes, P. (1991). A bottom-up mechanism for behavior selection in an artificial creature. In J.A. Meyer & S.W. Wilson (eds.), *From animals to animats* (pp. 238–246). Cambridge, MA: MIT Press.

Ritter, H., H. Cruse, & J. Dean (2000). *Prerational intelligence: Adaptive behavior and intelligent systems without symbols and logic,* Vol. 2. Dordrecht, The Netherlands: Kluwer Academic Publishers.

Searle, J.R. (1990). Is the brain's mind a computer program? *Scientific American* **262 (1)**, 26–31.

Tversky, A., & D. Kahneman (1986). The framing of decisions and the psychology of choice. In J. Elster (ed.), *Rational choice* (pp. 123–141). Oxford, UK: Blackwell.

GLOSSARY

Accommodation:
Creating a new psychological framework to incorporate new information (cf. assimilation).

Adaptive behavior:
1) Behavior furthering the reproductive fitness of the individual (biological fitness) or the goals and purposes of an animat; 2) adaptable; usefully changeable to improve utility or other performance measures.

Agent:
1) An entity which acts on and can be acted upon by its environment, e.g, autonomous agent; 2) in the sense of Minsky (1985) a largely independent neuronal unit (see also module).

Animat:
1) An artificial agent showing at least some quintessential properties of living beings; 2) an artificial, animal-like agent, e.g., an autonomous, mobile robot with at least a measure of autonomy; 3) an agent in an artificial life simulation.

Artificial life:
1) Computer and hardware simulations of quintessential processes of living systems, e.g., growth, development, morphogenesis, sociobiology, and evolution; 2) agents in such a simulation.

Assimilation:
Taking in new information within an existing psychological framework (cf. accommodation).

Associationist:
Conceiving of mental life as the association of ideas, without complex internal mental structure.

Autonomy:
1) Not dependent on another entity for energy or nutrients (energetic autonomy); 2) not under the control of another entity (motivational autonomy); 3) self-organizing, not dependent on another entity to guide its development or construction (developmental autonomy).

Basic behavior:
An elementary act from which more complex behaviors (i.e., compound behaviors) can be constructed.

Behavior:

1) The continuous stream of actions by an agent, animal or human, that can be registered by an observer; 2) a short-hand designation for categories of behavioral acts related according to function, form, perceived goal or some other characteristic.

Behavioral choice:

In neuroethology a mechanism – not necessarily a cognitive mechanism – for selecting from among different alternatives the behavior(s) to be performed at a given time.

Behavior-based architectures:

Robot control architectures rely on the combination of modules (or agents) that each control a different basic or compound behavior in order to generate more complex behaviors.

Behaviorists:

Psychologists of the early to mid-20th century who conceived of psychology as limited to the description of observable behavior, as opposed to internal mechanisms or mental processes, and therefore focused on describing links (S-R links) between stimuli (S) and responses (R).

Bottom-up:

1) influences or connections in a multi-layer network from lower levels, i.e., those closer to sensory inputs and presumably or demonstrably encoding simpler features, onto higher levels encoding more abstract and complex features; 2) design approach for starting with simple elements or functionality and combining these to achieve complex function or behavior (e.g., behavior-based architectures).

Chomskian language learning:

Learning language and especially grammar by choosing from among a set of genetically provided alternatives.

Coarse coding:

A means of encoding analog values with high precision by using a population of elements with overlapping domains where each element typically is capable of only low-precision encoding within its domain.

Cognition/cognitive:

1) Following Webster's Ninth New Collegiate Dictionary, the act or process of knowing including both awareness and judgment; 2) more generally, any process involving the application of available knowledge or the systematization of knowledge, either 2a) in a narrow sense, restricted to declarative and explicit knowledge (for explicit vs. implicit or easy vs. difficult to exploit, see *Mathematical* and *Philosophical Perspectives*), or 2b) in a broad sense including implicit knowledge (including procedural and perceptual processes, e.g., using prior explicit or implicit information to process and interpret current sensory inputs); 3) top-down influences; influences from higher levels – where more abstract and collective characteristics are represented – onto lower levels.

Cognitive psychology:
A psychology that studies internal functions – such as memory, decision-making, and other information-processing systems – that may intervene between stimulus and response (cf. behaviorists).

Collective intelligence:
Intelligence or adaptive behavior emerging through the interactions within a group of individuals, particularly when the behavior is beyond the competence of an isolated individual and therefore more than the simple sum of the contributions each individual could make on its own.

Constructivism:
A kind of boot-strapping by the individual: the theory that organisms develop progressively more powerful ways of interacting with the world in the course of interactions between internal and environmental influences.

Corollary discharge:
1) Originally defined as activity that is generated together with motor commands to muscles (the primary target) but travels to other centers, including the sender, whose inputs may be affected by the performance of the motor commands. Corollary discharges may increase or decrease the sensitivity or otherwise modify the activity of the receiver; 2) by extension, analogous recurrent or feedforward signals besides the primary output signals of a neural center.

Cost function:
An n to 1 function of variables representing the degrees of freedom in a system; the inverse of a utility function; as an added constraint, minimizing the cost function or maximizing the utility function is a way to specify a unique solution to problems with redundant degrees of freedom.

Credit assignment problem:
In optimization or learning the problem of determining which parameter changes affect the performance and which parameter changes are irrelevant.

Declarative knowledge:
Knowledge that can be readily expressed in words, what people can declare to be the case; also propositional knowledge, explicit knowledge, knowing that (see *Philosophical Perspectives*); neurological and behavioral data support distinguishing declarative knowledge from procedural knowledge.

Declarative memory:
"Knowing that", knowledge that can easily be expressed in words, factual knowledge; also manipulable knowledge.

Degrees of freedom:
The number of parameters necessary to completely and unambiguously specify the state of a system.

Deterministic chaos:
A particular form of progression of the state of a dynamical system; distinct from stochastic variation; characterized by a sensitive dependence on initial con-

ditions in the sense that two equivalent systems initially in closely neighboring states diverge exponentially over time, as opposed to converging to a common state or simply maintaining the original separation. A second required property is that of "mixing": The trajectories of distant states must be able to come close together again. This divergence might be considered to be destructive to order, but it may also serve to usefully magnify small differences.

Differentiation learning:
Learning to distinguish, that is, to treat differently, stimuli or situations previously treated the same to adapt a different, more differentiated strategy rather than sticking with fixed parameters in controlling behavior.

Dynamical or dynamic system:
A system for which the change in state over time can be written as a deterministic function of the current state, in analogy with systems studied in the branch of physics called mechanics; deterministic as opposed to stochastic; depending upon the equations the progression of the state of a dynamical system can be classified using the concepts of fixed points (either stable or unstable with respect to perturbations away from the fixed point), limit cycles, and deterministic chaos; the state space itself can be partitioned into basins of attraction for these structures; a variety of formal relationships can be established (see *Mathematical Perspectives*).

Dynamics:
Shorthand for the function describing the change of a dynamical system as a function of its current state.

Efference copy:
A corollary discharge containing a quantitatively faithful representation of the motor output; i.e., one that quantitatively relates to the normal effect of the efferent motor commands.

Emergent phenomena:
1) Properties of a collection of agents or elemental units that result from their interactions and require a new level of description beyond that adequate to describe the behavior of the individual agents; 2) properties not evident in the behavior of the individual units.

Empiricists:
Philosophers who consider experience the only source of knowledge and therefore historically emphasize the importance of environmental influences in development.

Epigenesis:
Development of the individual organism, (cf. phylogenesis, the evolution of species or other taxonomic groups).

Ethopsychology:
A psychology emphasizing the importance of the close interactions between the individual and its natural environment.

Evolution:
In the present context taken to mean not simply temporal change or historical development but rather Darwinian evolution, that is, change over time in the attributes of a population based upon the generation of heritable variation in individuals and differential reproduction based upon these variable attributes; imitated in artificial algorithms ("genetic algorithms").

Feature binding:
A mechanism for tagging activity arising in separate channels or modules as having a common referent; in particular, the linking of different characteristics (features) of a stimulus object analyzed in different parts of the brain.

Final common path:
Sherrington's description of motor neurons as the station through which all motor commands must pass, regardless of where they arise.

Flexibility:
Versatility and changeability evident in behavior, which may or may not be associated with plasticity in the control system.

Formal operations:
The final stage of cognitive development according to Piagetian psychology: A Piagetian stage in which a person can manipulate relations among ideas, deal with ideas known to be false, etc.

Forward modeling:
Computation of the effects of an action, (e.g., forward kinematics or dynamics: calculation of changes in the configuration of a multi-link body resulting from given changes in joint angles or joint torques, respectively), cf. inverse modeling.

Functional transition:
The process of transition to a new Piagetian stage (see "Piagetian psychology").

Hebbian learning:
A mechanism for learning based on increasing the synaptic strength between pre- and postsynaptic neurons when their activities are correlated, e.g., when activity in the presynaptic cell contributes to eliciting activity in the postsynaptic cell; initially hypothesized by Donald Hebb, the mechanism has been identified in biological nervous systems and implemented in many artificial neural networks.

Imprinting:
From ethology: the very early learning of simple responses, strongly guided by inborn tendencies and more pronounced during particular ages – so-called sensitive periods.

Information processing theory:
The approach that treats brain function as the processing of information in a computer-like manner.

Intelligence:
The capacity for intelligent behavior.

Intelligent behavior:
1) Behavior that is adaptive in the sense of furthering the reproduction of the biological individual and/or its genes or the survival of a technical agent, but with the implication of adaptability, the ability to deal with new situations; 2) a behavior incorporating to a greater degree some combination of attributes like autonomy, complexity, robustness, flexibility, generalization, learning, utility, and efficiency.

Intelligence of content:
An agent showing intelligent behavior by virtue of mechanisms relying on reasoning and symbolic thought.

Intelligence by design:
An agent showing intelligent behavior by virtue of processes, which may or may not include rational processes, that the agent possesses either as a result of evolution or design.

Intelligence of process:
An agent showing intelligent behavior by virtue of mechanisms that may (rational intelligence, intelligence of content) or may not (prerational intelligence) rely on explicitly rational processes and symbolic thought.

Intentionality:
Directed, adaptive selectivity in an agent's interactions with the environment and, by extension, in its procedures.

Inverse modeling:
Computation of the action needed to produce a specified effect, (e.g., inverse kinematics or dynamics: calculation of the joint changes or joint torques, respectively, needed to produce a desired change in the configuration of a multi-link body), cf. forward modeling.

Labeled line:
A particular neuron or channel, the activity of which is associated with a specific message; in contrast to a population code or across-fiber pattern code or coarse coding.

Learning:
A change in behavior resulting from interaction with the environment and not caused by development, injury or fatigue.

Maps:
A topographically ordered representation in a network, e.g., a representation of space, body parts or combinations thereof, other physical parameters or even non-physical parameters like semantic categories.

Module:
1) A cluster of elements in a network for which connections to other members of the cluster outweigh connections to non-members; 2) a unit within a larger entity

performing a separable function in a larger process, i.e., a functional unit; such a unit need not have a material counterpart, e.g., it can be a part of a computer program. The example of modular computer programs has led to the hypothesis of an analogous modular structure of the mind (Fodor 1983).

NP-complete:
A problem is NP-complete if the solution of any NP-hard decision problem can be reduced to this problem with an effort that grows at most polynomially with the problem size. It is believed (but not proven) that any NP-complete problem requires for its solution a computational effort that is at least exponential in the problem size.

Naive psychology:
The ideas of the average person about behavior and experience.

Neuromimetic:
A neural emulation of a structural capability.

O-calculus:
A formal approach to describing the order of complexity of a problem by determining upper boundaries as a function of the size of the input or argument. In particular, a function f on natural numbers is said to be of order $O(g)$ (or to belong to the class $O(g)$), if there exists a constant $C > 0$ such that $|f(n)| < |g(n)| \cdot C$ for all natural numbers $n > 0$. For example, $f(n) = 5n^2 + 17n - 35$ is in $O(n^2)$ with $g(n) = n^2$ and $C = 5 + 17$.

Optimization:
Finding the best solution from the space of possible solutions to a problem; besides the global optimum – the best solution in the space – there may also be local optima – solutions that are poorer than the global optimum but better than all their immediately neighboring solutions; regarding the quality of solutions as a landscape, the global optimum corresponds to the highest point while a local optimum corresponds to any hill top, no matter how low; optimization is generally a difficult, time-consuming problem (see *Computer Science Perspectives*); minimizing a cost function is equivalent to maximizing a utility function, but according to the landscape metaphor means finding low points; some algorithms (e.g., hill-climbing, gradient following methods) are deterministic and find only the local optimum for the particular starting point, whereas others (e.g., genetic algorithms, simulated annealing) can allow escape from the neighborhood of an unfavorable local optimum.

Parameter learning:
Learning that optimizes values of parameters in an algorithm.

Phase coupling:
Interactions between two or more rhythmic systems that depend only on the phases, i.e., on the phase difference, of the coupled systems, cf. pulse coupling.

Piagetian psychology:
The central tenet of Piaget and his followers stating that human development proceeds along the following stages: a sensorimotor stage of reflex organization, a preoperational stage of working with appearances, a concrete operational stage of linking appearances with representations, and a formal operational stage of manipulating ideas directly. Each has several substages.

Plan:
1) A representation of a goal that can motivate and control behavior to achieve the goal; a construct giving an agent a measure of independence from the immediate sensory inputs (see psychological perspectives); 2) a formal scheme for carrying out a sequence of actions such that a particular goal state is produced.

Plasticity:
1) structural change (e.g., change in weights or physical connections), which may be developmental or learned; 2) behavioral change, which may or may not depend on structural change.

Polysensory object:
An object apprehended simultaneously with several senses.

Population code:
A code in which the information is in the pattern of activity of a population of elements; across-fiber pattern or coarse coding are types of population codes; (cf. labeled line, where the presence or absence of activity in any single element carries a message).

Preoperational:
A Piagetian stage of development, where the child interacts with perceptions as they seem rather than with their internal representations.

Procedural knowledge/memory:
"Knowing how"; skills that cannot easily be expressed or communicated verbally and may even be subconscious; know-how, knowledge tied to a procedure; also implicit knowledge; cf. declarative knowledge.

Psychogenesis:
Development of psychological functions in an individual.

Pulse coupling:
Interactions between two or more rhythmic systems that are discontinuous, that is, occur only at restricted times within the cycle; systems with pulse coupling are analytically less tractable than those with phase coupling (see mathematical perspectives).

Rational:
Used here to mean a process explicitly involving symbolic thought and logical reasoning, as opposed to simply a reasonable course of action in a given circumstance; cf. prerational.

Reaction norm:
The range of normal developmental paths, in the absence of extraordinary influences.

Reactive agent:
An agent with no memory (e.g., no storage to hold a plan or a record of the past) whose behavior is driven solely by the current stimulus configuration.

Receptive field:
Set of locations in space where presentation of the adequate stimulus modifies, that is, increases or decreases, the activity of the sensory receptor or neuron being considered.

Recurrent pathway:
1) A pathway transmitting activity in the opposite direction to the predominant flow from sensory to motor elements; 2) a pathway in direction opposite to the dominant connection between two modules or layers; 3) a pathway within a network by which a change in the activity of a unit or group of units identifiably on the same layer can return to affect their inputs and subsequent activity (e.g., a feedback pathway), a collateral of an axon leaving a particular area which returns to the input region of the sender.

Reflex schemata:
Ways of responding to the immediate environment, usually early in development.

Reflexive intelligence:
Adaptive behavior in the form of a series of automatic fixed reactions to particular situations.

Representation:
1) A pattern of activity corresponding in an ordered way to a stimulus or action pattern, e.g., the direction of hand movement is represented in the population vector of activity in primary motor cortex, or the place cells in the hippocampus of mammals provide a representation of the animal's location and heading; 2) any encoding of aspects of stimulus, e.g., the environment, that is structurally different from the encoded original.

Schema/scheme (pl. schemata/schemes):
An internal algorithm for approaching a psychological problem, a representation of general specifications for categories of perception and action (e.g., a perceptual schema might contain lists of attributes for a particular class of objects, whereas a motor schema might contain movement primitives needing only variable parameters to be set in order to specify a specific action).

Second messenger:
A small, diffusible molecule serving as a signal within a cell – an intracellular messenger – that an extracellular messenger (e.g., hormone, neuromodulator, or neurotransmitter) has been received on the cell membrane; in some sensory receptors, second messengers are part of the transduction chain converting external stimulus to bioelectric signal in order to amplify the signal.

Self-determination:
Behavior controlled by internal plans rather than external demands.

Situated robot:
A robot considered explicitly in the context of a specific environment with which it interacts, as opposed to abstractly in terms of its general properties and capabilities.

Staging strategy:
In Piagetian psychology, a method for advancing to a new stage of development.

100 step rule:
A rough estimate of the number of processing steps between sensory input and voluntary response in humans based on typical reaction times and an assumed value of 10 ms for processing in an individual neuron; originally discussed in the context of visual perception and artificial vision (e.g., Ballard 1991), but used by extension as a general guideline for reasonable processing times in artificial systems.

Stigmergy:
After Grassé: communication via the environment as when the action of one animal modifies the environment so as to create or modify stimuli that change its own subsequent action and that of conspecifics (see *Biological Perspectives*).

Subsumption architecture:
A control architecture created by Rodney Brooks; it is a behavior-based architecture implementing specific logical relationships between the individual finite state control modules for the different behaviors.

Symbol:
After Webster's Ninth Collegiate Dictionary: something that stands for something else by reason of relationship, association or convention; an arbitrary or conventional sign.

Time complexity class NP:
The time complexity class NP includes most difficult optimization problems, e.g. the travelling salesman problem, scheduling problems, the knapsack problem, satisfiability problems of Boolean formulas, etc. A set A (formally, a subset of natural numbers) is said to be in NP when there exists a function f computable in polynomial time and a polynomial p such that the question of whether an element x belongs to A can be decided by the following procedure: $x \in A$ if and only if there exists a number w with length $|w|$ polynomially bounded by length $|x|$ (i.e., $|w| < p(|x|)$) and $f(w, x) = 1$. NP contains P (see time complexity class P); that NP is larger than P is commonly held to be true but as yet unproven.

Time complexity class P:
A set A is said to be in time complexity class P when the question of whether an element x belongs to A can be decided in polynomial time, that is, in a time bounded by a polynomial function of the length of the element x. Formally, the class P itself is defined on the natural numbers as the class of all sets A for which membership of x in A can be decided by computing a function $f(x)$

and the computation is bounded as stated. It is unknown whether P is a proper subset of or equivalent to NP. See also NP complete.

Top-down:
1) Influences or connections in a multi-layer network from higher levels, i.e., those more removed from sensory inputs and presumably or demonstrably encoding more abstract and complex features, back onto lower levels; 2) design principles beginning with a desired complex behavior or function and analytically decomposing it into simpler elements or functions.

Turing machine:
A minimalistic computing machine with the ability to read and write bits to a sequential tape (its external memory) and to move along the tape, a program in the form of a static decision table, and no internal memory (i.e., no state). Although cumbersome and more commonly employed as a theoretical concept than as a physical machine, it is as powerful as any computer language.

Utility:
The relationship of benefits to costs.

von Neumann architecture:
The conventional digital computer architecture as envisioned by John von Neumann; the central element of the hardware is a single active element, the central processor unit (CPU), that receives inputs, acts on data according to stored instructions (the program), produces outputs, and serves as a gatekeeper controlling the flow of data to and from memory. Particularly notable are the physical separation of active processing and passive memory and the associated restriction of processing to a small part of the hardware. Compare parallel distributed processing involving many active processing elements and integration of memory in the processing network.

Weight:
In an artificial neural network the parameter specifying the strength of the connection between two units, which determines how much the activity in the sender affects the activity in the receiver; equivalent to synaptic strength.

REFERENCES

Ballard, D. (1991) Animate vision. *Artificial Intelligence* **48**, 57–86.

Fodor, J. (1983). *The modularity of mind: An essay on faculty psychology.* Cambridge, MA: MIT Press.

LIST OF AUTHORS

Bauer, Hans-Ulrich Dr., Johann Wolfgang Goethe-Universität, Institut für Theoretische Physik, Robert-Mayer-Strasse 8-10, D-60054 Frankfurt/M., Germany

Beckers, Ralph, Dr., Université Libre de Bruxelles, Service de Chimie-Physique, Campus Plaine-CP, 231 Boulevard du Triomphe, B-1050 Bruxelles, Belgium, and Institut International de Recherche Betteravière, 47 Rue Montoyer, B-1090, Bruxelles, Belgium

Braun, Heinrich, Dr. habil., Systeme, Anwendungen und Produkte der Datenverarbeitung (SAP), Neurottstrasse 16, D-69189 Walldorf/Baden, Germany

Bridgeman, Bruce, Prof. Dr., Department of Psychology, University of California, Social Sciences 2, Santa Cruz, CA 95064, USA

Cellérier, Guy, Prof. Dr., Université de Genève, Faculté de Psychologie et des Sciences de l'Education, 40 boulevard du Pont-d'Arve, CH-1205 Genève, Switzerland

Cruse, Holk, Prof. Dr., Universität Bielefeld, Fakultät für Biologie, Postfach 100131, D-33501 Bielefeld, Germany

Dean, Jeffrey, Prof. Dr., Cleveland State University, BGES, 2399 Euclid Ave., Cleveland, OH 44115-2406, USA

Lanz, Peter†, Dr., Universität Bielefeld, Fakultät für Geschichtswissenschaft und Philosophie, Bielefeld, Germany

McFarland, David, Prof. Dr., Balliol College, Broad Street, Oxford 3BJ OX1, England

Paillard, Jacques, Prof. Dr., Centre National de la Recherche Scientifique – Neurobiologie et Mouvement, 31 chemin Joseph Aiguier, F-13402 Marseille Cedex 20, France

Ritter, Helge, Prof. Dr., Universität Bielefeld, Technische Fakultät, Postfach 100131, D-33501 Bielefeld, Germany

Velichkovsky, Boris M., Prof. Dr., Technische Universität Dresden, Fakultät Mathematik und Naturwissenschaften, Mommsenstrasse 13, D-01062 Dresden, Germany

SUBJECT INDEX

AUTHOR INDEX

307